Local Variance Estimation for Uncensored and Censored Observations

T0238480

Paola Gloria Ferrario

Local Variance Estimation for Uncensored and Censored Observations

Springer Vieweg

Paola Gloria Ferrario
Stuttgart, Germany

Dissertation University of Stuttgart, 2012

D 93

ISBN 978-3-658-02313-3 ISBN 978-3-658-02314-0 (eBook)
DOI 10.1007/978-3-658-02314-0

The Deutsche Nationalbibliothek lists this publication in the Deutsche Nationalbibliografie; detailed bibliographic data are available in the Internet at http://dnb.d-nb.de.

Library of Congress Control Number: 2013940101

Springer Vieweg
© Springer Fachmedien Wiesbaden 2013
This work is subject to copyright. All rights are reserved by the Publisher, whether the whole or part of the material is concerned, specifically the rights of translation, reprinting, reuse of illustrations, recitation, broadcasting, reproduction on microfilms or in any other physical way, and transmission or information storage and retrieval, electronic adaptation, computer software, or by similar or dissimilar methodology now known or hereafter developed. Exempted from this legal reservation are brief excerpts in connection with reviews or scholarly analysis or material supplied specifically for the purpose of being entered and executed on a computer system, for exclusive use by the purchaser of the work. Duplication of this publication or parts thereof is permitted only under the provisions of the Copyright Law of the Publisher's location, in its current version, and permission for use must always be obtained from Springer. Permissions for use may be obtained through RightsLink at the Copyright Clearance Center. Violations are liable to prosecution under the respective Copyright Law. The use of general descriptive names, registered names, trademarks, service marks, etc. in this publication does not imply, even in the absence of a specific statement, that such names are exempt from the relevant protective laws and regulations and therefore free for general use. While the advice and information in this book are believed to be true and accurate at the date of publication, neither the authors nor the editors nor the publisher can accept any legal responsibility for any errors or omissions that may be made. The publisher makes no warranty, express or implied, with respect to the material contained herein.

Printed on acid-free paper

Springer Vieweg is a brand of Springer DE.
Springer DE is part of Springer Science+Business Media.
www.springer-vieweg.de

"Das Wesen der Mathematik liegt in ihrer Freiheit"
Georg Cantor (1883)

(The essence of mathematics lies in its freedom)

Acknowledgements

At this point I would like to express my gratitude to Prof. Dr. em. Harro Walk for his excellent and intensive supervision and Dr. Maik Döring for precious advice. Many thanks also to Prof. Dr. Uwe Jensen, Prof. Dr. Ingo Steinwart, Prof. Maurizio Verri, Prof. Piercarlo Maggiolini and all the colleagues throughout the last years.
Last but certainly not least, I owe a special debt of gratitude to my husband for supporting me throughout the whole writing of this book.

Lübeck, April 2013 *Paola Gloria Ferrario*

Acknowledgements

[illegible]

[illegible]

Deutsche Zusammenfassung

Die mathematische Fragestellung, die in dieser Arbeit behandelt wird, hat eine mögliche Anwendung im medizinischen Bereich.
Wir nehmen an, dass ein Patient unter einer bestimmten Krankheit leidet und der behandelnde Arzt eine Prognose über den Krankheitsverlauf machen soll; insbesondere möchte er prognostizieren, ob nach der Heilung die Erkrankung wieder auftreten kann.
Dies kann einerseits aufgrund der Erfahrung des Arztes über den Verlauf der Krankheit geschehen. Anderseits stehen in den Krankenhäusern oft Datenbanken über Krankheitsverläufe von bereits behandelten Patienten zur Verfügung, die statistisch ausgewertet werden können.
Für jede Krankheit gibt es Faktoren oder Prädiktoren, die für den Verlauf der Krankheit oder das Wiederauftreten der Krankheit ausschlaggebend sind.
Von Interesse ist es, die mittlere Zeit bis zum nächsten Rückfall (die mittlere Überlebenszeit, die Rückfallwahrscheinlichkeit) $\mathbb{E}(Y \mid X = x) =: m(x)$ aufgrund einer Beobachtung des d-dimensionalen Prädiktor-Zufallsvektors X zu schätzen. Hier gibt die quadratisch integrierbare reelle Zufallsvariable Y die nicht beobachtbare Zeit bis zum nächsten Rückfall (die Überlebenszeit, den Rückfall-Indikatorwert) an. Die Vorhersagequalität der Regressionsfunktion $m : \mathbb{R}^d \to \mathbb{R}$ hängt wesentlich von den zur Verfügung stehenden Prädiktoren (z.B. Alter, Laborwerte, Anzahl der erlebten Rückfälle, Dosierung in den medikamentösen Therapien usw.) ab und wird global durch

$$\mathbb{E}((Y - m(X))^2$$

(sog. minimaler mittlerer quadratischer Fehler oder Residuenvarianz) und lokal durch

$$\sigma(x)^2 := \mathbb{E}((Y - m(X))^2 \mid X = x) = \mathbb{E}((Y^2 \mid X = x) - m(x)^2$$

(lokale Varianz) angegeben.
Ein Problem dabei ist, dass die Informationen, über die ein Krankenhaus verfügt, oft nicht vollständig sind oder, selbst während der Behandlung, aus verschiedenen Gründen enden (Zensierung).
Es wird angenommen, dass Y und die Zensierungszeit C unabhängig sind. Aufgrund der Zensierung liegen als Daten nicht Realisierungen von unabhängigen wie (X, Y) verteilten $(d+1)$-dimensionalen Zufallsvektoren (X_i, Y_i) vor (unzensierter Fall), sondern nur Realisierungen der unabhängigen wie (X, T, δ) verteilten Zufallsvektoren (X_i, T_i, δ_i) $(i = 1, ..., n)$, wobei $T = \min(Y, C)$, $\delta = 1_{\{Y \leq C\}}$.

Das Ziel dieser Arbeit ist die Schätzung der lokalen Varianz auf der Basis vorliegender Daten bezüglich X und Y - ohne und mit Zensierung - und die Untersuchung der asymptotischen Eigenschaften der Schätzung unter möglichst geringen Voraussetzungen an X und Y. Mehrere Schätzer der lokalen Varianz σ_n^2 anhand verschiedener Schätzmethoden werden angegeben, sowohl im unzensierten Fall als auch im zensierten Fall. Die Konsistenz dieser Schätzer wird gezeigt, und die entsprechende Konvergenzgeschwindigkeit unter Glattheitsvoraussetzungen wird ermittelt - in der auf die Verteilung von X bezogenen L_2- oder L_1-Norm.
Die Leistung zweier gewählter Schätzer wird durch Simulationen, auch im mehrdimensionalen Fall, untersucht.

Contents

List of Figures

List of Abbreviations

$(\Omega,\ \mathcal{A},\ P)$	Underlying probability space
i.i.d.	independent and identically distributed
$c,\ c_i,\ c^*,\ c^{**}\dots$	Constants
X	Random design (predictor) vector
Y	Dependent (response) random variable
$\boldsymbol{E}Y$	Expectation of Y
$\boldsymbol{Var}Y$	Variance of Y
$\overline{Y}$	Random variable Y contaminated by noise
C	Censoring time
T	$= \min\{Y, C\}$
$F_n(t)$	Kaplan-Meier estimator of the survival function $F(t) = \boldsymbol{P}(Y > t)$
$G_n(t)$	Kaplan-Meier estimator of the survival function $G(t) = \boldsymbol{P}(C > t)$
δ	$1_{\{Y \leq C\}}$
$\widetilde{Y} = \dfrac{\delta T}{G(T)}$	Censored random variable
$\widetilde{Y}_i = \dfrac{\delta_i T_i}{G_n(T_i)}$	Observations of the censored random variable under unknown survival function
$\widetilde{Y_i^2} = \dfrac{\delta_i T_i^2}{G_n(T_i)}$	Definition of $\widetilde{Y_i^2}$ $(\neq \widetilde{Y}_i^2)$

LS	Least Squares
LA	Local Averaging
1-NN, 2-NN, k-NN, NN	first, second, k-th nearest neighbors
$W_{n,i}(x; X_1, \ldots, X_n)$, $W_{n,i}(x)$	Weights, of partitioning, kernel or nearest neighbors type
m	Regression function
m_n	Regression function estimate
$m_n^{(LS)}$	Regression function estimate via Least Squares
$m_n^{(LA)}$	Regression function estimate via Local Averaging
$\overline{m}_n$	Regression function estimate under noise
$\widetilde{m}_n$	Regression function estimate under censoring
$Z := Y^2 - m^2(X)$	Random variable Z
$Z_i := Y_i^2 - m^2(X_i)$	Observations of the random variable Z under known m
$Z_i := Y_i^2 - m_n^2(X_i)$	Observations of the random variable Z under unknown m
$\overline{Z}_i := \overline{Y}_i^2 - \overline{m}_n^2(X_i)$	Observations of the random variable Z under noise
$\widetilde{Z}_i := \widetilde{Y_i^2} - \widetilde{m}_n^2(X_i)$	Observations of the random variable Z under censoring
σ^2	Local variance
$\sigma_n^{2\ (LS)}$	Local variance estimate via Least Squares
$\sigma_n^{2\ (LA)}$	Local variance estimate via Local Averaging
$\sigma_n^{2\ (NN)}$	Local variance estimate via 2-NN nearest neighbors
$\overline{\sigma}_n^2$	Local variance estimate under noise
$\widetilde{\sigma}_n^2$	Local variance estimate under censoring
$S_{K,M}$	Space of spline functions of order M with knot sequence K
$S_{K,M}^{L+1}([0,1]^d)$	Space of spline functions bounded by $L+1$
$S_{K,M}^{4L^2+1}([0,1]^d)$	Space of spline functions bounded by $4L^2+1$
a.s.	Almost sure
$\xrightarrow{P}$	Convergence in probability
O_P	Big O with respect to the probability measure P
$\|\ \|$	Euclidean norm

Conventions

$0/0 := 0.$

Ties occur with probability zero.

Definition

We call $S_n : \mathbb{R}^n \to \mathbb{R}^m$ stochastically bounded, if

$$\lim_{c\to\infty} \limsup_{n\to\infty} \boldsymbol{P}\left(\|S_n\| > c\right) = 0.$$

We write $S_n = O_P(c_n)$ if $(1/c_n)S_n$ is stochastically bounded.

Chapter 1
Introduction

Let X be a d-dimensional random vector (predictor vector) and Y be an integrable real random variable (dependent or response variable) on a probability space $(\Omega, \mathcal{A}, P)$.
The regression function $m : \mathbb{R}^d \to \mathbb{R}$ is defined by

$$m(x) := \boldsymbol{E}\{Y|X = x\}.$$

m allows to predict a non-observable realization of Y on the basis of an observed realization x of X by $m(x)$. In competition with other measurable functions $f : \mathbb{R}^d \to \mathbb{R}$ the expression $\boldsymbol{E}\{(Y - f(X))^2\}$ is minimized by m, i.e.,

$$V := \boldsymbol{E}\left\{(Y - m(X))^2\right\} = \min_f \boldsymbol{E}\left\{(Y - f(X))^2\right\}, \tag{1.1}$$

because of

$$\boldsymbol{E}\left\{|f(X) - Y|^2\right\} = \boldsymbol{E}\left\{(m(X) - Y)^2\right\} + \int |f(x) - m(x)|^2 \mu(dx)$$

where μ denotes the distribution P_X of X. V, the so-called residual variance (global variance), is a measure of how close we can get to Y using any measurable function f. It indicates how difficult a regression problem is.
Typically the distribution of (X, Y), and therefore m and V, are unknown. Nonparametric regression deals with the following problem: Given independent copies $(X_1, Y_1), \ldots, (X_n, Y_n)$ of (X, Y), an estimate m_n of the regression function shall be constructed, such that $\int |m_n(x) - m(x)|^2 \mu(dx)$ is "small".
The quality of predicting by the regression function m is locally given by the so-called local variance (or conditional variance)

$$\sigma^2(x) := \boldsymbol{E}\{(Y - m(X))^2|X = x\} = \boldsymbol{E}\{Y^2|X = x\} - m^2(x). \qquad (1.2)$$

The first aim of this thesis is to give estimators of the local variance σ_n^2 such that the squared L_2-distance $\int |\sigma_n^2(x) - \sigma^2(x)|^2 \mu(dx)$ is "small". Here μ denotes the distribution of X.

Comparing the literature, the majority of papers dealing with **nonparametric local variance estimation** concern fixed design. Here the one-dimensional case was treated by [23, 24, 12, 27, 1, 35] by using difference-based and plug-in methods. [13] and [26] treated the corresponding multi-dimensional case for a constant variance function (homoscedastic case) via differences, while in [30] under regularity assumptions via a local linear fit to the regression function optimal rate of convergence was obtained, also in the case of random design with density of X. Extending the work of [35, 1, 24, 23, 26], via differences of increasing order [2] investigated the general case of a non-constant (heteroscedastic) multivariate variance function.

The case of random design in estimating the heteroscedastic conditional variance was treated for $d = 1$ by [31] via recursions in view of σ^2 and m and by [14] in the context of time series. [9], also in the context of time series and with connection to fixed design, and [29] use local polynomial fitting for m and obtain local rates of convergence under higher differentiability assumptions. As [2], they pay attention to the adaptive situation where the rate of convergence is the same as if the true regression function would be known. [28] use wavelets in the case of a d-dimensional linear regression function. In the general framework of regression estimation for $d \geq 1$ with additional measurement errors for the $Y_i's$, [16] investigated as an application heteroscedastic conditional variance estimation (via plug-in) for $d = 1$ (for reasons of simplicity) by least squares methods (piecewise polynomial partitioning). Unlike the aforementioned authors, besides boundedness of X he imposes no conditions on the distribution of X, especially no density assumption. Under the same conditions as in [16], in **Chapter 2** we treat heteroscedastic local variance estimators of least squares type, show weak L_2-consistency and give a rate of convergence for sub-Gaussian $Y^2 - m^2(X)$, in the case of additional measurement errors in the dependent variable.

In **Chapter 3**, modifying Remark 5 in [16], we introduce local variance estimators of local averaging type, especially partitioning and kernel estimates, that are investigated in view of L_2-consistency and rate of convergence under $\boldsymbol{E}\{Y^4\} < \infty$. Also a version with splitting the sample is given. Furthermore the case of additional measurement errors in the dependent variables is treated.

Comparing again the literature, in [19], with generalization in [20], a nonparametric **estimator of the residual variance** V (1.1), introduced by [6, 7], was investigated, which is based on first and second nearest neighbors and the differences of the corresponding response variables. It simplifies an estimator given in [4] with the same convergence order, based on first nearest neighbors. **Chapter 4** deals with a local variance estimator which is a modification of the residual variance estimator of [19]. For this estimator, strong L_2-consistency under bounded Y, weak L_2-consistency under $\boldsymbol{E}\{Y^4\} < \infty$ and optimal rate of convergence under Hölder smoothness conditions are established.

The results of Chapters 2 to 4 may have an application in medical science. It is well known, that hospitals treating different patients collect data about them during the treatment of an illness. On the basis of these data the statistician helps physicians to predict whether the illness may come back after the treatment,e.g., predict the survival time of a new patient in dependence on different factors ("predictors"). A feature that complicates the analysis is that the follow-up program of the patients may be incomplete. After a certain censoring time, there is no information any longer about the patient.

The second aim of the thesis is to give estimators of the local variance also under censoring, using in addition the product-limit estimator, such that $\int |\sigma_n^2(x) - \sigma^2(x)|\mu(dx)$ is "small".

For that, let (X, Y, C), (X_1, Y_1, C_1), $(X_2, Y_2, C_2), \ldots$ be i.i.d. $\mathbb{R}^d \times \mathbb{R}_+ \times \mathbb{R}_+$-valued random vectors. X is the random vector of covariates with distribution μ, which, e.g., in medical applications contains information about a human taking part in a medical study around an illness. Y represents the survival time of the patient. C represents the censoring time. Moreover, we introduce the variable T, defined as minimum of Y and C, and the variable δ, containing the information whether there is or not censoring. This yields a set of data

$$\{(X_1, T_1, \delta_1), \ldots, (X_n, T_n, \delta_n)\},$$

with

$$\begin{cases} \delta_i = 1 \text{ for } Y_i \leq C_i \\ \delta_i = 0 \text{ for } Y_i > C_i, \end{cases}$$

and

$$T_i = \min\{Y_i, C_i\},$$

for $i = 1, \ldots, n$. In medical studies the observation of the survival time of the patient is sometimes incomplete due to RIGHT censoring formulated just before. It could, for example, happen that the patient is alive at the termination of a medical study, or that he dies by other causes than those

under study, or, trivially, that the patient moves and the hospital loses information about him.
Chapter 5 deals with local variance estimation for right censored observations. Analogously to Chapter 2, in Section 5.2 censored least squares estimates via plug-in are introduced. In Section 5.3 censored local averaging estimates via plug-in, analogously to Chapter 3, are treated and finally in Section 5.4 censored partitioning estimation via nearest neighbors, as in Chapter 4. For these estimators, weak L_1 consistency and rates are given.
Finally, **Chapter 6** deals with simulations of some proposed local variance estimators in the finite sample case for $d = 1$ as well as $d = 2$.

To facilitate access to the individual topics, the chapters are rendered as self-contained as possible.

1.1 Summary of the Main Results

For the convenience of the reader we repeat the relevant material from the following chapters without proofs, thus making our exposition self-contained.

Chapter 2 extends Kohler's results [16] and gives a least squares local variance estimator in the case of additional measurement errors in the dependent variable. The dataset is of the form

$$\overline{D}_n = \{(X_1, \overline{Y}_{1,n}), \dots, (X_n, \overline{Y}_{n,n})\},$$

where

$$\overline{Y}_i = m(X_i) + \overline{\epsilon}_i \quad \text{admits} \quad \boldsymbol{E}\{\overline{\epsilon}_i|X_i\} \neq 0,$$

in contrast to the common regression model where $\boldsymbol{E}\{\epsilon_i|X_i\} = 0$. Defining

$$Z := Y^2 - m^2(X),$$

we recognize that the local variance (1.2) is a regression on (X, Z). The observations of Z under noise and unknown m are given by

$$\overline{Z}_i := \overline{Y}_i^2 - \overline{m}_n^2(X_i)^{(LS)},$$

where

$$\overline{m}_n(\cdot)^{(LS)} = \arg\min_{f \in \mathcal{F}_n} \frac{1}{n} \sum_{i=1}^{n} |f(X_i) - \overline{Y}_i|^2,$$

$f : \mathbb{R}^d \to \mathbb{R} \in \mathcal{F}_n$, with suitable function space $\mathcal{F}_n$. Introduce therefore the following least squares local variance estimator under noise

$$\overline{\sigma}_n^2(\cdot)^{(LS)} = \arg\min_{g \in \mathcal{G}_n} \frac{1}{n} \sum_{i=1}^{n} |g(X_i) - \overline{Z}_i|^2, \tag{1.3}$$

where $g : \mathbb{R}^d \to \mathbb{R} \in \mathcal{G}_n$, with suitable function space $\mathcal{G}_n$ (so-called plug-in method because of use of $\overline{m}_n^{(LS)}$ instead of the unknown m for estimating $\sigma_n^{2\ (LS)}$).

The following theorem shows consistency of the least squares estimator of the local variance via plug-in (1.3).

Theorem 1.1. *Assume that $Y^2 - m^2(X)$ is sub-Gaussian in the sense that*

$$K^2 \boldsymbol{E}\left\{e^{(Y^2-m^2(X))^2/K^2} - 1|X\right\} \le \sigma_0^2 \quad \textit{almost surely}$$

for some K, $\sigma_0 > 0$. Let

$$\frac{1}{n}\sum_{i=1}^{n} |Y_i^p - \overline{Y}_i^p|^2 \xrightarrow{P} 0, \quad p = 1, 2.$$

It is assumed that $L^* > 0$ and $L > 0$ exist such that $\sigma^2 \le L^*$ and $|m| \le L$. Let $\mathcal{G}_n$ be defined as a subset of a linear space, consisting of nonnegative real-valued functions on $\mathbb{R}^d$ bounded by L^*, with dimension $D_n \in \mathbb{N}$, with the properties $\mathcal{G}_n \uparrow$, $D_n \to \infty$ for $n \to \infty$ but $\frac{D_n}{n} \to 0$. Furthermore $\cup_n \mathcal{G}_n$ is required to be dense in the subspace of $L_2(\mu)$ consisting of the nonnegative functions in $L_2(\mu)$ bounded by L^*. Let also $\mathcal{F}_n$ be defined as a subset of a linear space of real-valued functions on $\mathbb{R}^d$ absolutely bounded by L, with dimension $D'_n \in \mathbb{N}$, with the properties $\mathcal{F}_n \uparrow$, $D'_n \to \infty$ for $n \to \infty$ but $\frac{D'_n}{n} \to 0$. Furthermore $\cup_n \mathcal{F}_n$ is required to be a dense subset of $C^0_{0,L}(\mathbb{R}^d)$ (with respect to the max norm), where $C^0_{0,L}(\mathbb{R}^d)$ denotes the space of continuous real valued functions on $\mathbb{R}^d$ absoluted bounded by L, with compact support. Then

$$\int |\overline{\sigma}_n^2(x)^{(LS)} - \sigma^2(x)|^2 \mu(dx) \xrightarrow{P} 0.$$

The following theorem gives a convergence rate of the estimator (1.3). Here $S_{K,M}^{L+1}([0,1]^d)$ and $S_{K,M}^{4L^2+1}([0,1]^d)$ are the following spline spaces

$$S_{K,M}^{L+1}([0,1]^d) := \left\{ \sum_i \alpha_i B_{i,K,M}^d : 0 \le \alpha_j \le L+1 \right\}$$

and

$$S_{K,M}^{4L^2+1}([0,1]^d) := \left\{ \sum_i \alpha_i B_{i,K,M}^d : 0 \le \alpha_j \le 4L^2+1 \right\}$$

($j \in \{1, \ldots, K+M\}^d$) where $B_{i,K,M}^d$ is a multivariate B-spline of order M with knot sequence K.

Theorem 1.2. *Let $L \ge 1$, $C > 0$ and $p = k + \beta$ for some $k \in \mathbb{N}_0$ and $\beta \in (0,1]$. Assume that $X \in [0,1]^d$ almost surely. Assume also that $|Y_i| \le L$, $|\overline{Y}_i| \le L$ and*

$$\frac{1}{n}\sum_{i=1}^{n} |Y_i - \overline{Y}_i|^2 = O_P\left(n^{-\frac{2p}{2p+d}}\right).$$

Moreover, let $\Gamma > 0$, $\Lambda > 0$ and assume that m and σ^2 are (p, Γ) and (p, Λ)-smooth, respectively, that is, for every $\alpha = (\alpha_1, \ldots, \alpha_d)$, $\alpha_j \in \mathbb{N}_0$, $\sum_{j=1}^{d} \alpha_j = k$

$$\left| \frac{\partial^k m}{\partial x_1^{\alpha_1}, \ldots, \partial x_d^{\alpha_d}}(x) - \frac{\partial^k m}{\partial x_1^{\alpha_1}, \ldots, \partial x_d^{\alpha_d}}(z) \right| \le \Gamma \|x - z\|^\beta \quad x,\ z \in \mathbb{R}^d$$

and

$$\left| \frac{\partial^k \sigma^2}{\partial x_1^{\alpha_1}, \ldots, \partial x_d^{\alpha_d}}(x) - \frac{\partial^k \sigma^2}{\partial x_1^{\alpha_1}, \ldots, \partial x_d^{\alpha_d}}(z) \right| \le \Lambda \|x - z\|^\beta \quad x,\ z \in \mathbb{R}^d$$

($\| \ \|$ denoting the Euclidean norm). Identify $\mathcal{F}_n$ and $\mathcal{G}_n$ with $S_{K_n',M}^{L+1}([0,1]^d)$ and $S_{K_n,M}^{4L^2+1}([0,1]^d)$, respectively, with respect to an equidistant partition of $[0,1]^d$ into

$$K_n' = \lceil \Gamma^{\frac{2}{2p+d}} n^{\frac{1}{2p+d}} \rceil$$

for $\mathcal{F}_n$ and

$$K_n = \lceil \Lambda^{\frac{2}{2p+d}} n^{\frac{1}{2p+d}} \rceil,$$

for $\mathcal{G}_n$, respectively. Then

$$\int \left| \overline{\sigma}_n^2(x)^{(LS)} - \sigma^2(x) \right|^2 \mu(dx) = O_P\left(n^{-\frac{2p}{2p+d}}\right).$$

In **Chapter 3**, modifying Kohler's Remark 5 in [16], we introduce local variance estimators of local averaging type, especially partitioning and kernel estimates. The estimation of m is given by the weighted mean of those Y_i where X_i is in a certain sense close to x :

$$m_n(x)^{(LA)} = m_n^{(LA)}(x, X_1, Y_1, \ldots, X_n, Y_n) = \sum_{i=1}^{n} W_{n,i}(x) \cdot Y_i, \tag{1.4}$$

where the weights $W_{n,i}(x, X_1, \ldots, X_n) \in \mathbb{R}$, briefly written as $W_{n,i}(x)$, depend on $X_1, \ldots, X_n$ and are therefore nondeterministic. We have "small" (nonnegative) weights in the case that X_i is "far" from x. Kernel, partitioning and nearest neighbors weights are widespread (see, for instance, [11]). By $Z = Y^2 - m^2(X)$ and observations

$$Z_{n,i} := Y_i^2 - m_n^2(X_i)^{(LA)},$$

recognize that the local variance (1.2) is a regression on (X, Z).
The proposed local averaging estimator of the local variance is

$$\sigma_n^2(x)^{(LA)} = \sum_{i=1}^{n} W_{n,i}(x) \cdot Z_{n,i}, \tag{1.5}$$

in dependence on the different weights (again plug-in method because of use of the estimate $m_n^{(LA)}$ of m for estimating $\sigma_n^{2\ (LA)}$). Note that $W_{n,i} = W_{n,i}^{(1)}$ in (1.5) and $W_{n,i} = W_{n,i}^{(2)}$ in (1.4).

$$W_{n,i}(x, X_1, \ldots, X_n) = \frac{1_{A_n(x)}(X_i)}{\sum_{l=1}^{n} 1_{A_n(x)}(X_l)} \tag{1.6}$$

are partitioning weights, with $0/0 := 0$, where $\mathcal{P}_n = \{A_{n,1}, A_{n,2}, \ldots\}$ is a partition of $\mathbb{R}^d$ consisting of Borel sets $A_{n,j} \subset \mathbb{R}^d$, and where the notation $A_n(x)$ is used for the $A_{n,j}$ containing x. Further kernel weights are known in the literature, depending on the kernel $K : \mathbb{R}^d \to [0, \infty)$,

$$W_{n,i}(x, X_1, \ldots, X_n) = \frac{K\left(\frac{x - X_i}{h_n}\right)}{\sum_{l=1}^{n} K\left(\frac{x - X_l}{h_n}\right)}. \tag{1.7}$$

There, let $h_n > 0$ be the bandwidth and $0/0 := 0$ again. Common kernels are for example the naive kernel ($K(x) = 1_{\{\|x\|<1\}}$) and the Epanechnikov kernel ($K(x) = (1 - \|x\|^2)_+$). For our aims we introduce now the so-called boxed kernel. Define as $S_{0,r}$ the balls of Radius r and $S_{0,R}$ the balls of radius R, both centered in 0, $0 < r \leq R$. Let b be a positive constant. The symmetric boxed kernel fulfills the properties

$$1_{\{x \in S_{0,R}\}} \geq K(x) \geq b 1_{\{x \in S_{0,r}\}}.$$

Finally, nearest neighbor weights are also frequently used, defined by

$$W_{n,i}(x, X_1, \ldots, X_n) = \frac{1}{k_n} 1_{\{X_i \text{ is among the } k_n \text{ NNs of } x \text{ in } \{X_1, \ldots, X_n\}\}} \tag{1.8}$$

($2 \leq k_n \leq n$), here usually assuming that ties occur with probability 0.
We distinguish local averaging methods in the auxiliary estimates m_n (1.4) and in the estimates σ_n^2 in (1.5), indicating the weights by $W_{n,i}^{(2)}$ and $W_{n,i}^{(1)}$ (instead of $W_{n,i}$ in (1.5)), respectively. Thus

$$m_n(X_i) = \sum_{j=1}^{n} W_{n,j}^{(2)}(X_i, X_1, \ldots, X_n) Y_j$$

where

$$W_{n,j}^{(2)}(x, X_1, \ldots, X_n),$$

is of partitioning type, with partitioning sequence $\left\{A_{n,j}^{(2)}\right\}$, or of kernel type, with kernel $K^{(2)}$ and bandwidths $h_n^{(2)}$, or of nearest neighbor type, with $k_n^{(2)}$ neighbors.
The following theorem show consistency of the local averaging estimator of the local variance via plug-in (1.5).

Theorem 1.3. *Let (X, Y) have an arbitrary distribution with $\mathbf{E}\{Y^4\} < \infty$. For partitioning weights defined according to (1.6) assume that, for each sphere S centered at the origin*

$$\lim_{n \to \infty} \max_{j:\ A_{n,j}^{(l)} \cap S \neq \emptyset} diam(A_{n,j}^{(l)}) = 0, \quad l = 1, 2,$$

$$\lim_{n \to \infty} \frac{|\{j : A_{n,j}^{(l)} \cap S \neq \emptyset\}|}{n} = 0, \quad l = 1, 2.$$

For kernel weights defined according to (1.7) with kernels $K^{(l)}$ assume that the bandwidths satisfy

$$0 < h_n^{(l)} \to 0, \quad n h_n^{(l)d} \to \infty, \quad l = 1, 2,$$

($K^{(l)}$ symmetric, $1_{S_{0,R}}(x) \geq K^{(l)}(x) \geq b 1_{S_{0,r}}(x)$ $(0 < r \leq R < \infty,\ b > 0)$). For nearest neighbor weights defined according to (1.8) assume that

$$2 \leq k_n^{(2)} \leq n, \quad k_n^{(2)} \to \infty, \quad \frac{k_n^{(2)}}{n} \to 0$$

Then for the estimate (1.5) under the above assumptions

$$\lim_{n\to\infty} \mathbf{E} \int \left| \sigma_n^2(x)^{(LA)} - \sigma^2(x) \right| \mu(dx) = 0$$

holds.

The following theorem gives a rate of convergence of the estimator (1.5).

Theorem 1.4. *Let the estimate of the local variance σ^2 be given by (1.5) with weights $W_{n,i}^{(1)}(x)$ of cubic partition with side length $h_n^{(1)}$ or with naive kernel $1_{S_{0,1}^{(1)}}$ with bandwidths $h_n^{(1)}$, further for $m_n(X_i)$ given by (1.4) with weights $W_{n,i}^{(2)}(x)$ of cubic partition with side length $h_n^{(2)}$ or with naive kernel $1_{S_{0,1}}$ and bandwidths $h_n^{(2)}$ or with $k_n^{(2)}$-nearest neighbors (the latter for $d \geq 2$). Assume that X is bounded and that*

$$|Y| \leq L \in [0, \infty),$$

$$|m(x) - m(z)| \leq C\|x - z\|, \quad x, z \in \mathbb{R}^d,$$

and finally, that

$$|\sigma^2(x) - \sigma^2(z)| \leq D\|x - z\|, \quad x, z \in \mathbb{R}^d$$

($\|\ \|$ denoting the Euclidean norm). Then, for

$$h_n^{(1)} \sim n^{-\frac{1}{d+2}},$$

and

$$h_n^{(2)} \sim n^{-\frac{1}{d+2}}, \quad \text{and} \quad k_n^{(2)} \sim n^{\frac{2}{d+2}}, \quad \text{respectively,}$$

$$\mathbf{E} \int |\sigma_n^2(x)^{(LA)} - \sigma^2(x)| \mu(dx) = O\left(n^{-\frac{1}{d+2}}\right).$$

In **Chapter 4** we introduce a partitioning local variance estimator based on the first and second nearest neighbors. The k-th nearest neighbor of X_i

among $X_1, \ldots, X_{i-1}, X_{i+1}, \ldots, X_n$ is defined as $X_{N[i,k]}$ with

$$N[i,k] := N_n[i,k] := \underset{1 \le j \le n,\ j \ne i,\ j \notin \{N[i,1], \ldots, N[i,k-1]\}}{\arg\min} \rho(X_i, X_j).$$

Here ρ is a metric (typically the Euclidean one) in $\mathbb{R}^d$.
Our proposal for an appropriate estimator of σ^2 is

$$\sigma_n^2(x)^{(NN)} := \frac{\sum_{i=1}^n (Y_i - Y_{N[i,1]})(Y_i - Y_{N[i,2]}) 1_{A_n(x)}(X_i)}{\sum_{i=1}^n 1_{A_n(x)}(X_i)}, \quad x \in \mathbb{R}^d \tag{1.9}$$

with $0/0 := 0$, where $\mathcal{P}_n = \{A_{n,1}, A_{n,2}, \ldots\}$ is a partition of $\mathbb{R}^d$ consisting of Borel sets $A_{n,j} \subset \mathbb{R}^d$, and where the notation $A_n(x)$ is used for the $A_{n,j}$ containing x. Ties shall occur with probability zero.
The following theorem shows strong consistency of the partitioning estimator of the local variance via nearest neighbors (1.9).

Theorem 1.5. *Let $(\mathcal{P}_n)_{n \in \mathbb{N}}$ with $\mathcal{P}_n = \{A_{n,1}, A_{n,2}, \ldots\}$ be a sequence of partitions of $\mathbb{R}^d$ such that for each sphere S centered at the origin*

$$\lim_{n \to \infty} \max_{j : A_{n,j} \cap S \neq \emptyset} \mathit{diam}\, A_{n,j} \to 0$$

and, for some $\rho = \rho(S) \in (0, \frac{1}{2})$,

$$\#\{j : A_{n,j} \cap S \neq \emptyset\} \sim n^\rho.$$

Assume $|Y| \le L$ for some $L \in \mathbb{R}_+$. Then

$$\int |\sigma_n^2(x)^{(NN)} - \sigma^2(x)|^2 \mu(dx) \to 0 \quad a.s.$$

The following theorem gives a rate of the estimator (1.9).

Theorem 1.6. *Assume that X is bounded and*

$$\boldsymbol{E}\{Y^4 | X = x\} \le \tau^4, \quad x \in \mathbb{R}^d,$$

$(0 < \tau < \infty)$. Moreover, assume the Hölder conditions

$$|\sigma^2(x) - \sigma^2(t)| \le D \|x - t\|^\beta, \quad x, t \in \mathbb{R}^d,$$

and

$$|m(x) - m(t)| \le D^* \|x - t\|^\alpha, \quad x, t \in \mathbb{R}^d,$$

with $0 < \alpha \le 1$, $0 < \beta \le 1$; C, $D \in \mathbb{R}_+$, ($\|\ \|$ denoting the Euclidean norm). Let $\mathcal{P}_n$ be a cubic partition of $\mathbb{R}^d$ with side length h_n of the cubes ($n \in \mathbb{N}$).

Then, with

$$h_n \sim n^{-\frac{1}{2\beta+d}}$$

for the estimate (1.9) one has

$$\boldsymbol{E}\left\{\int |\sigma_n^2(x)^{(NN)} - \sigma^2(x)|^2 \mu(dx)\right\} = O\left(\max\left\{n^{-4\alpha/d}, n^{-2\beta/(2\beta+d)}\right\}\right).$$

Only uniform boundedness of the conditional fourth moment of Y and boundedness of X, especially no density condition on X, are assumed. Thus in the case of random design, for Hölder continuity the partitioning variance estimate of simple structure with first and second nearest neighbors (without plug-in) yields the same convergence order as in [2] for regular fixed design. If $4\alpha/d \geq 2\beta/(2\beta+d)$, i.e., $\alpha \geq \beta d/(2(2\beta+d))$, one has the same convergence rate as in the case of known m, i.e., in the case of classic partitioning regression estimation with dependent variable $(Y - m(X))^2$. For the class of Hölder continuous functions σ^2 with exponent $\beta \leq 1$ this convergence rate $n^{-2\beta/(2\beta+d)}$ is optimal see p. 37, Theorem 3.2, p. 66, Theorem 4.3 with proof in [11], i.e., the sequence $n^{-2\beta/(2\beta+d)}$ is the lower minimax rate and is achieved, namely by partitioning estimates.

Chapter 5 deals with local variance estimation under censoring.
From (X, Y, C), (X_1, Y_1, C_1), $(X_2, Y_2, C_2), \ldots$ i.i.d. $\mathbb{R}^d \times \mathbb{R}_+ \times \mathbb{R}_+$-valued random vectors, the modified underlying set of data is of type

$$\{(X_1, T_1, \delta_1), \ldots, (X_n, T_n, \delta_n)\},$$

with

$$\begin{cases} \delta_i = 1 \text{ for } Y_i \leq C_i \\ \delta_i = 0 \text{ for } Y_i > C_i, \end{cases}$$

and

$$T_i = \min\{Y_i, C_i\},$$

for $i = 1, \ldots, n$. We introduce now the so-called survival functions

$$F(t) = \boldsymbol{P}(Y > t),$$

$$G(t) = \boldsymbol{P}(C > t),$$

and

$$K(t) = \boldsymbol{P}(T > t) = F(t)G(t),$$

$T := \min\{Y, C\}$. The survival functions map the event of survival onto time and are therefore monotone decreasing. Set

$$T_F := \sup\{y : \ F(y) > 0\},$$

$$T_G := \sup\{y : \ G(y) > 0\},$$

$$T_K := \sup\{y : \ K(y) > 0\} = \min\{T_F, T_G\}.$$

For our intents we require the following conditions, which are required throughout the whole chapter 5:

(A1) C and (X, Y) are independent,
(A2) $\exists L > 0$, such that $\boldsymbol{P}\{\max\{Y, Y^2\} \leq L\} = 1$ and $\boldsymbol{P}\{C > L\} > 0$.
G is continuous.
(A3) $\forall 0 < T'_K < T_K : \ \boldsymbol{P}\{0 \leq Y \leq T'_K\} < 1, \ \boldsymbol{P}\{0 \leq Y^2 \leq T'_K\} < 1.$
F is continuous in a neighborhood of T_K and in a neighborhood of $\sqrt{T_K}$.

Let F_n and G_n be the Kaplan-Meier estimates of F and G, respectively, which are defined by

$$F_n(t) = \begin{cases} \prod_{i=1,\ldots,n \ T(i) \leq t} \left(\frac{n-i}{n-i+1}\right)^{\delta_{(i)}} & t \leq T(n) \\ 0 & \text{otherwise} \end{cases}$$

and

$$G_n(t) = \begin{cases} \prod_{i=1,\ldots,n \ T(i) \leq t} \left(\frac{n-i}{n-i+1}\right)^{1-\delta_{(i)}} & t \leq T(n) \\ 0 & \text{otherwise,} \end{cases}$$

where $((T(1), \delta(1)), \ldots, (T(n), \delta(n)))$ are the n pairs of observed (T_i, δ_i) set in increasing order.
Set then

$$\widetilde{Y^2} := \frac{\delta T^2}{G(T)}$$

and their observations (G is known)

$$\widetilde{Y_i^2} = \frac{\delta_i T_i^2}{G(T_i)},$$

and, for unknown G,

$$\widetilde{Y_{n,i}^2} = \frac{\delta_i T_i^2}{G_n(T_i)}.$$

Finally, set

$$\widetilde{Z}_{n,i} := \widetilde{Y^2_{n,i}} - \widetilde{m}^2_n(X_i)^{(LS)}$$

with

$$\widetilde{m}_n(\cdot)^{(LS)} := \arg\min_{f \in \mathcal{F}_n} \frac{1}{n} \sum_{i=1}^{n} |f(X_i) - \widetilde{Y}_i|^2,$$

where $f : \mathbb{R}^d \to \mathbb{R} \in \mathcal{F}_n$, $\mathcal{F}_n$ being a suitable function space.
The least squares estimator of the local variance under censoring introduced in section 5.2 is

$$\widetilde{\sigma}^2_n(\cdot)^{(LS)} := \arg\min_{g \in \mathcal{G}_n} \frac{1}{n} \sum_{i=1}^{n} |g(X_i) - \widetilde{Z}_i|^2, \tag{1.10}$$

($g : \mathbb{R}^d \to \mathbb{R} \in \mathcal{G}_n$, $\mathcal{G}_n$ being a suitable function space).
The following theorem shows consistency of the least squares estimator of the local variance via plug-in, under censoring (1.10). It is analogous to Theorem 1.1 for the uncensored case.

Theorem 1.7. *Assumptions* ***(A1)-(A2)*** *hold and moreover $X \in [0,1]^d$. Let $\mathcal{G}_n$ be defined as a subset of a linear space, consisting of nonnegative real-valued functions on $\mathbb{R}^d$ bounded by L^*, with dimension $D_n \in \mathbb{N}$, with the properties $\mathcal{G}_n \uparrow$, $D_n \to \infty$ for $n \to \infty$ but $\frac{D_n}{n} \to 0$. Furthermore $\cup_n \mathcal{G}_n$ is required to be dense in the subspace of $L_2(\mu)$ consisting of the nonnegative functions in $L_2(\mu)$ bounded by L^*. Let also $\mathcal{F}_n$ be defined as a subset of a linear space of real-valued functions on $\mathbb{R}^d$ absolutely bounded by L, with dimension $D'_n \in \mathbb{N}$, with the properties $\mathcal{F}_n \uparrow$, $D'_n \to \infty$ for $n \to \infty$ but $\frac{D'_n}{n} \to 0$. Furthermore $\cup_n \mathcal{F}_n$ is required to be a dense subset of $C^0_{0,L}(\mathbb{R}^d)$ (with respect to the max norm), where $C^0_{0,L}(\mathbb{R}^d)$ denotes the space of continuous real valued functions on $\mathbb{R}^d$ absoluted bounded by L, with compact support. Then*

$$\int |\widetilde{\sigma}^2_n(x)^{(LS)} - \sigma^2(x)|^2 \mu(dx) \xrightarrow{P} 0. \tag{1.11}$$

The following theorem gives a rate of convergence of a modification of the estimator (1.10), consisting of a truncation of $\widetilde{m}_n$ and $\widetilde{\sigma}^2_n$ at height L.

Theorem 1.8. *Assumptions* ***(A1)-(A3)*** *hold. $X \in [0,1]^d$ almost surely. Moreover, let $\Gamma > 0$, $\Lambda > 0$ and $p = k + \beta$ for some $k \in \mathbb{N}_0$ and $\beta \in (0,1]$. m and σ^2 are (p, Γ) and (p, Λ)-smooth, respectively, that is, for every $\alpha = (\alpha_1, \ldots, \alpha_d)$, $\alpha_j \in \mathbb{N}_0$, $\sum_{j=1}^d \alpha_j = k$*

$$\left| \frac{\partial^k m}{\partial x_1^{\alpha_1}, \ldots, \partial x_d^{\alpha_d}}(x) - \frac{\partial^k m}{\partial x_1^{\alpha_1}, \ldots, \partial x_d^{\alpha_d}}(z) \right| \le \Gamma \|x - z\|^\beta \quad x,\ z \in \mathbb{R}^d$$

and

$$\left| \frac{\partial^k \sigma^2}{\partial x_1^{\alpha_1}, \ldots, \partial x_d^{\alpha_d}}(x) - \frac{\partial^k \sigma^2}{\partial x_1^{\alpha_1}, \ldots, \partial x_d^{\alpha_d}}(z) \right| \leq \Lambda \|x - z\|^\beta \quad x,\ z \in \mathbb{R}^d$$

($\| \|$ denoting the Euclidean norm).
Identify $\mathcal{F}_n$ and $\mathcal{G}_n$ with $S_{K'_n,M}([0,1]^d)$ and $S_{K_n,M}([0,1]^d)$, respectively, with respect to an equidistant partition of $[0,1]^d$ into

$$K'_n = \lceil \Gamma^{\frac{2}{2p+d}} \left(\frac{\log n}{n} \right)^{\frac{1}{2p+d}} \rceil$$

for $\mathcal{F}_n$ and

$$K_n = \lceil \Lambda^{\frac{2}{2p+d}} \left(\frac{\log n}{n} \right)^{\frac{1}{2p+d}} \rceil,$$

for $\mathcal{G}_n$, respectively.
Then, with truncation of $\widetilde{m}_n$ and $\widetilde{\sigma}_n^2$ at height L (without changing the notation),

$$\int \left| \widetilde{\sigma}_n^2(x) - \sigma^2(x) \right|^2 \mu(dx) = O_P \left(\left(\frac{\log n}{n} \right)^{\frac{1}{3}} + \left(\frac{\log n}{n} \right)^{\frac{2p}{2p+d}} \right).$$

In Section 5.3, analogously to Chapter 3, local variance estimates of local averaging type are introduced, for the censored case.
For that, by

$$\widetilde{Z}_{n,i} = \widetilde{Y_{n,i}^2} - \widetilde{m}_n^2(X_i)^{(LA)},$$

with censored regression estimator

$$\widetilde{m}_n(x)^{(LA)} := \sum_{i=1}^n W_{n,i}(x) \widetilde{Y}_{n,i},$$

the local variance estimator (plug-in method) is defined by

$$\widetilde{\sigma}_n^2(x)^{(LA)} := \sum_{i=1}^n W_{n,i}(x) \widetilde{Z}_{n,i}. \tag{1.12}$$

The following theorem shows consistency of the local averaging estimator of the local variance via plug-in, under censoring (1.12). It is analogous to Theorem 1.3 for the uncensored case.

Theorem 1.9. *Let the assumptions **(A1)** and **(A2)** hold. For partitioning weights defined according to (1.6) assume that, for each sphere S centered at the origin*

$$\lim_{n\to\infty} \max_{A_{n,j}^{(l)}\cap S\neq\emptyset} \operatorname{diam}(A_{n,j}^{(l)}) = 0, \quad l = 1,2,$$

$$\lim_{n\to\infty} \frac{|\{j : A_{n,j}^{(l)} \cap S \neq \emptyset\}|}{n} = 0, \quad l = 1,2.$$

For kernel weights defined according to (1.7) with kernels $K^{(l)}$ assume that the bandwidths satisfy

$$0 < h_n^{(l)} \to 0, \quad nh_n^{(l)d} \to \infty, \quad l = 1,2,$$

($K^{(l)}$ symmetric, $1_{S_{0,R}}(x) \geq K^{(l)}(x) \geq b1_{S_{0,r}}(x)$ $(0 < r \leq R < \infty,\ b > 0)$). For nearest neighbor weights defined according to (1.8), which refer only to $\widetilde{m}_n^{(LA)}(x)$ assume that

$$2 \leq k_n^{(2)} \leq n, \quad k_n^{(2)} \to \infty, \quad \frac{k_n^{(2)}}{n} \to 0$$

Then for the estimate (1.12) under the above assumptions

$$\int \left|\widetilde{\sigma}_n^2(x)^{(LA)} - \sigma^2(x)\right| \mu(dx) \xrightarrow{P} 0$$

holds.

The following theorem gives a rate of convergence of the estimator (1.12).

Theorem 1.10. *Let the assumptions **(A1)-(A3)** hold. Let the estimate $\sigma_n^{2\ (LA)}$ be given by (1.12) with weights $W_{n,i}$ as in (1.7) and naive kernel $1_{\{S_{0,1}\}}$ with bandwidth $h_n \sim n^{-\frac{1}{d+2}}$. Moreover let m and σ^2 be Lipschitz continuous, that is*

$$|m(x) - m(z)| \leq \Gamma\|x - z\| \quad x,\ z \in \mathbb{R}^d$$

and

$$|\sigma^2(x) - \sigma^2(z)| \leq \Lambda\|x - z\| \quad x,\ z \in \mathbb{R}^d,$$

($\Lambda,\ \Gamma \in \mathbb{R}_+$, $\|\ \|$ denoting the Euclidean norm).

Then

$$\int |\widetilde{\sigma}_n^2(x)^{(LA)} - \sigma^2(x)|\mu(dx) = O_P\left(\left(\frac{\log n}{n}\right)^{\frac{1}{6}} + n^{-\frac{1}{d+2}}\right).$$

Finally, analogously to Chapter 4, in Section 5.4 we introduce partitioning estimates via nearest neighbors under censoring. For that, introduce

$$\widetilde{\sigma}_n^2(x)^{(NN)} := \frac{\sum_{i=1}^n H_{i,G_n} 1_{A_n(x)}(X_i)}{\sum_{i=1}^n 1_{A_n(x)}(X_i)} \tag{1.13}$$

where

$$\begin{aligned} H_{i,G_n} &:= H_{n,i,G_n} \\ &= \frac{\delta_i T_i^2}{G_n(T_i)} - \frac{\delta_i T_i}{G_n(T_i)} \frac{\delta_{N[i,1]} T_{N[i,1]}}{G_n(T_{N[i,1]})} - \frac{\delta_i T_i}{G_n(T_i)} \frac{\delta_{N[i,2]} T_{N[i,2]}}{G_n(T_{N[i,2]})} \\ &\quad + \frac{\delta_{N[i,1]} T_{N[i,1]}}{G_n(T_{N[i,1]})} \frac{\delta_{N[i,2]} T_{N[i,2]}}{G_n(T_{N[i,2]})} \end{aligned}$$

The following theorem shows consistency of the partitioning estimator of the local variance via nearest neighbors under censoring and unknown survival function.

Theorem 1.11. *Let Assumptions **(A1)-(A3)** hold. Let $\mathcal{P}_n = \{\boldsymbol{A}_{n,1}, \dots, \boldsymbol{A}_{n,l_n}\}$ be a sequence of partitions on $\mathbb{R}^d$ such that for each sphere S centered at the origin*

$$\lim_{n\to\infty} \max_{j\in\{A_{n,j}\cap S\neq\emptyset\}} \operatorname{diam} \boldsymbol{A}_{n,j} = 0,$$

and

$$\lim_{n\to\infty} \frac{\#\{j:\ A_{n,j} \cap S \neq \emptyset\}}{n} = 0.$$

Then

$$\int |\widetilde{\sigma}_n^2(x)^{(NN)} - \sigma^2(x)|\mu(dx) \xrightarrow{P} 0.$$

The following theorem gives a convergence rate of estimator (1.13).

Theorem 1.12. *Let the assumptions **(A1)-(A3)** hold. Let the estimate $\widetilde{\sigma}^{2\ (NN)}$ be given by (1.13) with cubic partition of $\mathbb{R}^d$ with side length h_n of the cubes ($n \in \mathbb{N}$). Moreover, assume the Lipschitz conditions*

$$|m(x) - m(t)| \leq \Gamma \|x - t\|^\alpha,\ x, t \in \mathbb{R}^d,$$

and

$$|\sigma^2(x) - \sigma^2(t)| \leq \Lambda \|x - t\|^\beta, \ x, t \in \mathbb{R}^d,$$

($0 < \alpha \leq 1$, $0 < \beta \leq 1$, Γ, $\Lambda \in \mathbb{R}_+$, $\| \ \|$ denoting the Euclidean norm). Then, with

$$h_n \sim n^{-\frac{1}{d+2\beta}}$$

one gets

$$\int |\widetilde{\sigma}_n^{2\ (NN)} - \sigma^2(x)| \mu(dx) = O_P\left(\left(\frac{\log n}{n}\right)^{\frac{1}{6}} + \max\left\{n^{-\frac{2\alpha}{d}}, n^{-\frac{\beta}{2\beta+d}}\right\}\right).$$

Chapter 2
Least Squares Estimation via Plug-In

2.1 Regression Estimation

Under i.i.d. random vectors (X, Y), (X_1, Y_1), $(X_2, Y_2), \ldots$ in the regression analysis one is interested in the value of the so called response variable Y (in $\mathbb{R}$) depending on the value of the observation vector X (in $\mathbb{R}^d$, with distribution μ). To find that, one searches a (measurable) function $f : \mathbb{R}^d \to \mathbb{R}$, such that $f(X)$ is a "good approximation of Y", that is, $f(X)$ should be "close" to Y, achieved making the random quantity $|f(X) - Y|$ "small". For this, assuming square integrability of Y, we introduce the so-called L_2-risk or mean squared error of f:

$$\boldsymbol{E}\left\{|f(X) - Y|^2\right\}. \tag{2.1}$$

It is well known that the function that minimizes in a certain sense (2.1) is the regression function, $m(x) := \boldsymbol{E}\{Y|X = x\}$, unknown if the distribution of (X, Y) is unknown. In this case, starting from a dataset D_n, a nonparametric estimator m_n of the regression function is to construct. There exist different paradigms how to make it, the aim here is to deal with least squares approaches. There, the basic idea is to estimate the unknown mean squared error in (2.1) by approximating the expectation value there appearing via the empirical mean:

$$\frac{1}{n}\sum_{i=1}^{n} |f(X_i) - Y_i|^2, \tag{2.2}$$

and to choose a function, over a set $\mathcal{F}_n$ of functions given by the statistician, $f : \mathbb{R}^d \to \mathbb{R}$, that minimizes (2.2).
Examples of possible choices of the set $\mathcal{F}_n$ are sets of piecewise polynomials with respect to a partition $\mathcal{P}_n$. Clearly it doesn't make sense to minimize

(2.2) over all measurable functions f, because this may lead to a function which interpolates the data and hence is not a reasonable estimate.
The least squares estimator of the regression function is defined as:

$$m_n(\cdot) = \arg\min_{f \in \mathcal{F}_n} \frac{1}{n} \sum_{i=1}^{n} |f(X_i) - Y_i|^2,$$

where the optimal function is not required to be unique. For consistency and rate of convergence of such estimators see [16] and the references cited there.
Sometimes it is possible to observe data from the underlying distribution only with measurement errors. It can for example happen, that the predictor vector X can be observed only with errors, i.e., instead of X_i one observes $W_i = X_i + U_i$ for some random variable U_i which satisfy $\boldsymbol{E}\{U_i|X_i\} = 0$ and the aim is to estimate the regression function from $\{(W_1, Y_1), \ldots, (W_n, Y_n)\}$. Instead, as in [16], we assume that we can observe the dependent variable Y only with supplementary, maybe correlated, measurement errors. Since we do not assume that the means of these measurement errors are zero, these kinds of errors are not already included in standard models. Our dataset is

$$\overline{D}_n = \{(X_1, \overline{Y}_{1,n}), \ldots, (X_n, \overline{Y}_{n,n})\},$$

where the only assumption on the random variables $\overline{Y}_{1,n}, \ldots, \overline{Y}_{n,n}$ is that the differences between Y_i and $\overline{Y}_{i,n}$ are in a certain sense "small". We will therefore assume that the average squared measurement error

$$\frac{1}{n} \sum_{i=1}^{n} |Y_i - \overline{Y}_{i,n}|^2$$

is small. Set briefly $\overline{Y}_i := \overline{Y}_{i,n}$. With the difficulty of additional measurement errors in the dependent variable in our notation the estimator becomes:

$$\overline{m}_n(\cdot)^{(LS)} = \arg\min_{f \in \mathcal{F}_n} \frac{1}{n} \sum_{i=1}^{n} |f(X_i) - \overline{Y}_i|^2, \tag{2.3}$$

and in this chapter we set $\overline{m}_n(x) := \overline{m}_n(x)^{(LS)}$.
Inspired by [16], Corollary 1, we want here to treat consistency in a general case where we require the unknown regression function only to be bounded in absolute value from above by a constant and without continuity assumptions.

Theorem 2.1. *Assume that $Y - m(X)$ is sub-Gaussian in the sense that*

$$K^2 \boldsymbol{E}\left\{e^{(Y-m(X))^2/K^2} - 1|X\right\} \leq \sigma_0^2 \quad \textit{almost surely,}$$

for some K, $\sigma_0 > 0$. Let $L \geq 1$ and assume that the regression function is bounded in absolute value by L ($\Rightarrow m \in L_2(\mu)$). We define $\mathcal{F}_n$ as a subset of a linear space, consisting of real-valued functions on $\mathbb{R}^d$, with dimension $D_n \in \mathbb{N}$ and with the property $|f| \leq L$ for $f \in \mathcal{F}_n$, where $\mathcal{F}_n \uparrow$, $D_n \to \infty$ for $n \to \infty$, but $\frac{D_n}{n} \to 0$. Furthermore $\cup_n \mathcal{F}_n$ is required to be dense in the subspace of $L_2(\mu)$, consisting of the functions in $L_2(\mu)$ absolutely bounded by L. In addition it shall hold

$$\frac{1}{n} \sum_{i=1}^{n} |Y_i - \overline{Y}_i|^2 \xrightarrow{P} 0. \tag{2.4}$$

Then, we have

$$\int |\overline{m}_n(x) - m(x)|^2 \mu(dx) \xrightarrow{P} 0.$$

(Consistency of the least squares estimator of the regression function with additional measurements error in the response variable)

For the proof of Theorem 2.1 we use the following lemma:

Lemma 2.1. *$\{U_n\}$ and $\{V_n\}$ are nonnegative real random sequences. Assume $V_n \xrightarrow{P} 0$ and $\boldsymbol{P}\{U_n > V_n\} \to 0$ $(n \to \infty)$. Then $U_n \xrightarrow{P} 0$.*

Proof. For each $\varepsilon > 0$ and each $\delta > 0$ there exists an n_0 such that, for every $n \geq n_0$:

$$\boldsymbol{P}\underbrace{\{V_n > \varepsilon\}}_{\Omega_n'} \leq \frac{\delta}{2}$$

and

$$\boldsymbol{P}\underbrace{\{U_n > V_n\}}_{\Omega_n''} \leq \frac{\delta}{2},$$

thus

$$\begin{aligned}
&\boldsymbol{P}(\{U_n \leq \varepsilon\}^c) \leq \boldsymbol{P}(\{\{V_n \leq \varepsilon\} \cap \{U_n \leq V_n\}\}^c) \\
&= \boldsymbol{P}(\{\Omega_n'^c \cap \Omega_n''^c\}^c) \overset{\text{De Morgan}}{=} \boldsymbol{P}(\{\Omega_n' \cup \Omega_n''\}) \\
&\leq \boldsymbol{P}\{\Omega_n'\} + \boldsymbol{P}\{\Omega_n''\} \leq \frac{\delta}{2} + \frac{\delta}{2} = \delta.
\end{aligned}$$

□

We are now ready for the following:

Proof of Theorem 2.1. According to [16], Corollary 1, there is a positive constant c depending only on L, σ_0, K with the following property:

$$\boldsymbol{P}\left\{\int |\overline{m}_n(x) - m(x)|^2 \mu(dx) \right.$$
$$\left. > c\left(\frac{1}{n}\sum_{i=1}^{n} |Y_i - \overline{Y}_i|^2 + \frac{D_n}{n} + \inf_{f \in \mathcal{F}_n} \int |f(x) - m(x)|^2 \mu(dx)\right)\right\} \to 0.$$

By Lemma 2.1, it is enough to show

$$\left(\frac{1}{n}\sum_{i=1}^{n} |Y_i - \overline{Y}_i|^2 + \frac{D_n}{n} + \inf_{f \in \mathcal{F}_n} \int |f(x) - m(x)|^2 \mu(dx)\right) \xrightarrow{P} 0 \tag{2.5}$$

But (2.5) holds, because of (2.4), $\frac{D_n}{n} \to 0$ and $\inf_{f \in \mathcal{F}_n} \int |f(x) - m(x)|^2 \mu(dx) \to 0$ (due to $\mathcal{F}_n \uparrow$ and the density of $\cup_n \mathcal{F}_n$ in the mentioned subspace of $L_2(\mu)$).

2.2 Local Variance Estimation with Additional Measurement Errors

The quality of the regression function m in view of small mean squared error is globally given by $\boldsymbol{E}\{(Y - m(X))^2\}$ and locally by

$$\sigma^2(x) := \boldsymbol{E}\{(Y - m(X))^2 | X = x\} = \boldsymbol{E}\{Y^2 | X = x\} - m^2(x). \tag{2.6}$$

$\sigma^2(x)$ is the so called local variance. We define a new variable

$$Z := Y^2 - m^2(X) \tag{2.7}$$

and consequently its observations (in the case of known m):

$$Z_i := Y_i^2 - m^2(X_i);$$

finally the observations with additional errors:

$$\overline{Z}_i := \overline{Y}_i^2 - \overline{m}_n^2(X_i),$$

with $\overline{m}_n = \overline{m}_n^{(LS)}$ according to (2.4). Combining (2.6) and (2.7) allows us to say that the local variance is a regression on (X, Z).

We can therefore define the least squares estimator of the local variance, analogously to the estimator (2.3) as

$$\overline{\sigma}_n^2(\cdot)^{(LS)} = \arg\min_{g \in \mathcal{G}_n} \frac{1}{n} \sum_{i=1}^{n} |g(X_i) - \overline{Z}_i|^2, \tag{2.8}$$

where $g : \mathbb{R}^d \to \mathbb{R} \in \mathcal{G}_n$, with suitable function space $\mathcal{G}_n$. Briefly, define $\overline{\sigma}_n^2(x) := \overline{\sigma}_n^2(x)^{(LS)}$.
For consistency and convergence rate under Lipschitz conditions see [16]. We want here to treat consistency in a general case where no smoothness conditions on the m and σ^2 are required.

Theorem 2.2. *Assume that $Y^2 - m^2(X)$ is sub-Gaussian in the sense that*

$$K^2 \mathbf{E}\left\{ e^{(Y^2 - m^2(X))^2/K^2} - 1 | X \right\} \leq \sigma_0^2 \quad \textit{almost surely}$$

for some K, $\sigma_0 > 0$. Let

$$\frac{1}{n} \sum_{i=1}^{n} |Y_i^p - \overline{Y}_i^p|^2 \xrightarrow{P} 0, \quad p = 1, 2. \tag{2.9}$$

It is assumed that $L^* > 0$ and $L > 0$ exist such that $\sigma^2 \leq L^*$ and $|m| \leq L$. Let $\mathcal{G}_n$ be defined as a subset of a linear space, consisting of nonnegative real-valued functions on $\mathbb{R}^d$ bounded by L^*, with dimension $D_n \in \mathbb{N}$, with the properties $\mathcal{G}_n \uparrow$, $D_n \to \infty$ for $n \to \infty$ but $\frac{D_n}{n} \to 0$. Furthermore $\cup_n \mathcal{G}_n$ is required to be dense in the subspace of $L_2(\mu)$ consisting of the nonnegative functions in $L_2(\mu)$ bounded by L^*. Let also $\mathcal{F}_n$ be defined as a subset of a linear space of real-valued functions on $\mathbb{R}^d$ absolutely bounded by L, with dimension $D'_n \in \mathbb{N}$, with the properties $\mathcal{F}_n \uparrow$, $D'_n \to \infty$ for $n \to \infty$ but $\frac{D'_n}{n} \to 0$. Furthermore $\cup_n \mathcal{F}_n$ is required to be a dense subset of $C^0_{0,L}(\mathbb{R}^d)$ (with respect to the max norm), where $C^0_{0,L}(\mathbb{R}^d)$ denotes the space of continuous real valued functions on $\mathbb{R}^d$ absoluted bounded by L, with compact support. Then

$$\int |\overline{\sigma}_n^2(x) - \sigma^2(x)|^2 \mu(dx) \xrightarrow{P} 0.$$

(Consistency of the least squares estimator of the local variance with additional measurements error in the response variable)

Remark 2.1. In case of boundedness of Y_i and $\overline{Y}_i$ (2.9) for $p = 1$ implies (2.9) for $p = 2$ (because of $Y_i^2 - \overline{Y}_i^2 = (Y_i - \overline{Y}_i)(Y_i + \overline{Y}_i)$).

Proof of Theorem 2.2. As in the proof of Theorem 2.1 we obtain that there exists a generic positive constant c depending only from L, σ_0, K with the following property:

$$\boldsymbol{P}\left\{\int |\overline{\sigma}_n^2(x) - \sigma^2(x)|^2 \mu(dx) > c \cdot \left(\frac{1}{n}\sum_{i=1}^n |Z_i - \overline{Z_i}|^2 + \frac{D_n}{n} + \inf_{g \in \mathcal{G}_n} \int |g(x) - \sigma^2(x)|^2 \mu(dx)\right)\right\} \to 0. \tag{2.10}$$

We notice

$$\begin{aligned}
&\frac{1}{n}\sum_{i=1}^n |Z_i - \overline{Z}_i|^2 \\
&\le \frac{2}{n}\sum_{i=1}^n |\overline{m}_n^2(X_i) - m^2(X_i)|^2 + \frac{2}{n}\sum_{i=1}^n |Y_i^2 - \overline{Y}_i^2|^2 \\
&\le \frac{8}{n}L^2 \sum_{i=1}^n |\overline{m}_n(X_i) - m(X_i)|^2 + \underbrace{\frac{2}{n}\sum_{i=1}^n |Y_i^2 - \overline{Y}_i^2|^2}_{\stackrel{P}{\to}0 \quad \text{assumption (2.9)}}.
\end{aligned} \tag{2.11}$$

It remains to prove

$$\frac{1}{n}\sum_{i=1}^n |\overline{m}_n(X_i) - m(X_i)|^2 \stackrel{P}{\to} 0.$$

Via conditioning with respect to $(X_1, \dots, X_n)$, by [16], Lemma 3, we obtain

$$\begin{aligned}
&\boldsymbol{P}\Bigg\{\frac{1}{n}\sum_{i=1}^n |\overline{m}_n(X_i) - m(X_i)|^2 \\
&> c\left(\frac{1}{n}\sum_{i=1}^n |Y_i - \overline{Y_i}|^2 + \frac{D_n'}{n} \right.\\
&\quad \left. + \min_{f \in \mathcal{F}_n} \frac{1}{n}\sum_{i=1}^n |f(X_i) - m(X_i)|^2\right) \Bigg| X_1 = x_1, \dots, X_n = x_n\Bigg\} \to 0
\end{aligned} \tag{2.12}$$

where $\frac{D_n'}{n} \to 0$ and $\frac{1}{n}\sum_{i=1}^n |Y_i - \overline{Y}_i|^2 \stackrel{P}{\to} 0$ (assumption (2.9)).
Regarding the last term in the round brackets in (2.12), for an arbitrary

we choose $\varepsilon' > 0$ a continuous function with compact support $\tilde{m}$ such that $\boldsymbol{E}|\tilde{m}(X) - m(X)|^2 \leq \varepsilon'$.
We observe

$$\min_{f \in \mathcal{F}_n} \frac{1}{n} \sum_{i=1}^{n} |f(X_i) - m(X_i)|^2 \leq$$

$$\leq 2 \underbrace{\min_{f \in \mathcal{F}_n} \frac{1}{n} \sum_{i=1}^{n} |f(X_i) - \tilde{m}(X_i)|^2}_{\to 0 \ a.s.} \tag{2.13}$$

$$+ 2 \underbrace{\frac{1}{n} \sum_{i=1}^{n} |\tilde{m}(X_i) - m(X_i)|^2}_{a.s. \ \to \boldsymbol{E}\{|\tilde{m}(X) - m(X)|^2\} \leq \varepsilon' \text{ (Strong Law of Large Numbers)}}$$

where (2.13) follows from

$$\min_{f \in \mathcal{F}_n} \frac{1}{n} \sum_{i=1}^{n} |f(X_i) - \tilde{m}(X_i)|^2$$

$$\leq \min_{f \in \mathcal{F}_n} \left[\sup_x |f(x) - \tilde{m}(x)|^2 \right] \to 0.$$

By Lemma 2.1 the assertion follows.

2.3 Rate of Convergence

In this section we investigate the rate of convergence of the least squares estimator of the local variance function with additional measurement errors in the dependent variable. In the special case that there are no additional measurement errors and that $d = 1$ Kohler's Corollary 3 [16] investigates the rate of convergence of the estimator.
In this section we choose as suitable function space for the minimization problem in (2.8) the space of B-spline functions, as Mathe did [22].
We recall now briefly the definitions of B-splines and the B-splines space. For a deeper discussion of splines with proofs we refer the reader to the classical reference here [3].

Definition 2.1. Let $K := (K_i)$ be a nondecreasing sequence. The i-th univariate (normalized) B-spline of order M for the knot sequence K is denoted by the rule

$$B_{i,K,M}(x) := (K_{i+M} - K_i)[K_i, \ldots, K_{i+M}](K - x)_+^{M-1} \quad \text{for } x \in \mathbb{R}.$$

Notice that B-splines consists of nonnegative functions which sum up to 1, i.e., $B_{i,K,M}$ provides a partition of unity. Further explanations can be found again in [3].

Definition 2.2. For $i = (i_1, \ldots, i_d) \in \mathbb{Z}^d$, the multivariate B-splines of order m are denoted by

$$B^d_{i,K,M}(x_1, \ldots, x_d) := B_{i_1,K,M}(x_1) \cdot \ldots \cdot B_{i_d,K,M}(x_d)$$

Definition 2.3. A spline function of order M with knot sequence K is any linear combination of B-splines of order M for the knot sequence K. The collection of all such functions is denoted by $S_{K,M}$. In symbols,

$$S_{K,M}([0,1]^d) := \left\{ \sum_i \alpha_i B^d_{i,K,M} : \alpha_i \text{ real, for all } i \right\}.$$

Notice that the functions from $S_{K,M}$ are multivariate polynomials of degree smaller or equal to M and for $M > 0$ they are $(M-1)$-times continuously differentiable.
Because of the bound of the regression function and the local variance function it makes sense to bound also the functions of the spline space. Therefore, we bound the estimate introducing the following two modifications of the spline space $S_{K,M}$

$$S^{L+1}_{K,M}([0,1]^d) := \left\{ \sum_i \alpha_i B^d_{i,K,M} : 0 \le \alpha_j \le L+1 \right.$$
$$\left. (j \in \{1, \ldots, K_n + M\}^d) \right\}$$

and

$$S^{4L^2+1}_{K,M}([0,1]^d) := \left\{ \sum_i \alpha_i B^d_{i,K,M} : 0 \le \alpha_j \le 4L^2+1 \right.$$
$$\left. (j \in \{1, \ldots, K'_n + M\}^d) \right\}.$$

Because of the properties of the B-splines to be positive and to sum up to one, the functions from the space $S^{L+1}_{K,M}([0,1]^d)$ are nonnegative and bounded by $L+1$. Analogously, the functions from $S^{4L^2+1}_{K,M}([0,1]^d)$ are bounded by

$4L^2 + 1$. The following theorem deals with the rate of convergence of the estimator of the local variance.

Theorem 2.3. *Let $L \geq 1$, $C > 0$ and $p = k + \beta$ for some $k \in \mathbb{N}_0$ and $\beta \in (0,1]$. Assume that $X \in [0,1]^d$ almost surely. Assume also that $|Y_i| \leq L$, $|\overline{Y}_i| \leq L$ and*

$$\frac{1}{n}\sum_{i=1}^{n}|Y_i - \overline{Y}_i|^2 = O_P\left(n^{-\frac{2p}{2p+d}}\right). \tag{2.14}$$

Moreover, let $\Gamma > 0$, $\Lambda > 0$ and assume that m and σ^2 are (p,Γ) and (p,Λ)-smooth, respectively, that is, for every $\alpha = (\alpha_1, \ldots, \alpha_d)$, $\alpha_j \in \mathbb{N}_0$, $\sum_{j=1}^{d}\alpha_j = k$

$$\left|\frac{\partial^k m}{\partial x_1^{\alpha_1}, \ldots, \partial x_d^{\alpha_d}}(x) - \frac{\partial^k m}{\partial x_1^{\alpha_1}, \ldots, \partial x_d^{\alpha_d}}(z)\right| \leq \Gamma \|x - z\|^{\beta} \quad x,\ z \in \mathbb{R}^d$$

and

$$\left|\frac{\partial^k \sigma^2}{\partial x_1^{\alpha_1}, \ldots, \partial x_d^{\alpha_d}}(x) - \frac{\partial^k \sigma^2}{\partial x_1^{\alpha_1}, \ldots, \partial x_d^{\alpha_d}}(z)\right| \leq \Lambda \|x - z\|^{\beta} \quad x,\ z \in \mathbb{R}^d$$

($\| \ \|$ denoting the Euclidean norm).
Identify $\mathcal{F}_n$ and $\mathcal{G}_n$ with $S^{L+1}_{K_n',M}([0,1]^d)$ and $S^{4L^2+1}_{K_n,M}([0,1]^d)$, respectively, with respect to an equidistant partition of $[0,1]^d$ into

$$K_n' = \lceil \Gamma^{\frac{2}{2p+d}} n^{\frac{1}{2p+d}} \rceil$$

for $\mathcal{F}_n$ and

$$K_n = \lceil \Lambda^{\frac{2}{2p+d}} n^{\frac{1}{2p+d}} \rceil,$$

for $\mathcal{G}_n$, respectively. Then

$$\int \left|\overline{\sigma}_n^2(x) - \sigma^2(x)\right|^2 \mu(dx) = O_P\left(n^{-\frac{2p}{2p+d}}\right).$$

(Rate of convergence of the least squares estimator of the local variance with additional measurements error in the response variable)

Proof. We use (2.10). Because of the dimension $D_n = c \cdot K_n$ of $\mathcal{G}_n$ it follows

$$\frac{D_n}{n} \leq O\left(n^{-\frac{2p}{2p+d}}\right). \tag{2.15}$$

From the (p, Γ)-smoothness of σ^2 and the definition of $\mathcal{G}_n$ we can conclude (cf. [22], p. 66)

$$\inf_{g \in \mathcal{G}_n} \int |g(x) - \sigma^2(x)|^2 \mu(dx) \leq O\left(n^{-\frac{2p}{2p+d}}\right) \tag{2.16}$$

In view of the assertion it remains to show

$$\frac{1}{n}\sum_{i=1}^{n} \left|Z_i - \overline{Z}_i\right|^2 = O_P\left(\Lambda^{\frac{1}{2p+d}} n^{-\frac{p}{2p+d}}\right).$$

Now we use (2.11). It holds

$$\frac{1}{n}\sum_{i=1}^{n} \left|Y_i^2 - \overline{Y}_i^2\right|^2 = O_P\left(\Lambda^{\frac{1}{2p+d}} n^{-\frac{p}{2p+d}}\right)$$

because of

$$\frac{1}{n}\sum_{i=1}^{n} |Y_i^2 - \overline{Y}_i^2|^2 \leq 4L^2 \cdot \frac{1}{n}\sum_{i=1}^{n} \left|Y_i - \overline{Y}_i\right|^2$$

by uniform boundedness of the sequence $\frac{1}{n}\left(\sum_{i=1}^{n} \left|Y_i + \overline{Y}_i\right|^2\right)^{1/2}$, and (2.14). Thus it remains to show

$$\frac{1}{n}\sum_{i=1}^{n} |\overline{m}_n(X_i) - m(X_i)|^2 = O_P\left(\Lambda^{\frac{2}{2p+d}} n^{-\frac{2p}{2p+d}}\right).$$

We work now conditionally on $(X_1, \ldots, X_n)$ and observe that

$$\begin{aligned}
&\boldsymbol{P}\Biggl\{\frac{1}{n}\sum_{i=1}^{n} |\overline{m}_n(X_i) - m(X_i)|^2 \\
&> c\left(\frac{1}{n}\sum_{i=1}^{n} |Y_i - \overline{Y}_i|^2 + \frac{D_n'}{n}\right. \\
&\quad \left. + \min_{f \in \mathcal{F}_n} \frac{1}{n}\sum_{i=1}^{n} |f(X_i) - m(X_i)|^2\right) \Bigg| X_1 = x_1, \ldots, X_n = x_n\Biggr\} \to 0
\end{aligned} \tag{2.17}$$

where $\frac{D'_n}{n} \leq O\left(n^{-\frac{2p}{2p+d}}\right)$, analogously to (2.15) and
$\min_{f\in\mathcal{F}_n} \frac{1}{n}\sum_{i=1}^{n}|f(x_i) - m(x_i)|^2 \leq O\left(n^{-\frac{2p}{2p+d}}\right)$, analogously to (2.16).
This, together with (2.14), implies (2.17) and therefore the assertion. □

Chapter 3
Local Averaging Estimation via Plug-In

3.1 Local Variance Estimation with Splitting the Sample

Let (X, Y), (X_1, Y_1), $(X_2, Y_2), \ldots$ be independent and identically distributed $\mathbb{R}^d \times \mathbb{R}$-valued random vectors with $|Y| \leq L \in [0, \infty)$. Here no additive noise for Y is assumed. The regression function $m : \mathbb{R}^d \to \mathbb{R}$ is defined by

$$m(x) := \boldsymbol{E}\{Y|X = x\}.$$

The regression function m allows to predict a non-observable realization of Y on the basis of an observed realization x of X by $m(x)$, where in competition with other functions $f : \mathbb{R}^d \to \mathbb{R}$ the mean squared error $\boldsymbol{E}\{(Y - m(X))^2\}$ is minimal. m is unknown if the distribution of (X, Y) is unknown. Given $(X_1, Y_1), \ldots, (X_n, Y_n)$, an estimate $m_n(x)$ of the regression function shall be constructed, such that $\int |m_n(x) - m(x)|^2 \mu(dx)$ is "small" (μ denoting the distribution P_X of X). For more details on regression estimation see [11]. By the local averaging method the estimation of $m(x)$ is given by the weighted mean of those Y_i where X_i is in a certain sense close to x :

$$m_n(x)^{(LA)} = m_n^{(LA)}(x, X_1, Y_1, \ldots, X_n, Y_n) = \sum_{i=1}^{n} W_{n,i}(x) \cdot Y_i, \qquad (3.1)$$

where the weights $W_{n,i}(x, X_1, \ldots, X_n) \in \mathbb{R}$, briefly written as $W_{n,i}(x)$, depend on $X_1, \ldots, X_n$ and are therefore nondeterministic. We have "small" (nonnegative) weights in the case that X_i is "far" from x. Depending on the definition of the weights, we distinguish between partitioning, kernel and nearest neighbor estimates.

The quality of predicting by the regression function m is locally given by the local variance

$$\sigma^2(x) := \boldsymbol{E}\{(Y - m(X))^2 | X = x\} = \boldsymbol{E}\{Y^2 | X = x\} - m^2(x).$$

We define now new random variables

$$Z := Y^2 - m^2(X)$$

and in context of observations in the case of known regression function

$$Z_i := Y_i^2 - m^2(X_i)$$

and of unknown regression function

$$Z_{n,i} := Y_i^2 - m_n^2(X_i).$$

Notice that usually m is unknown and has to be estimated. In this way one has a plug-in method. Recognize that the local variance function is a regression on the pair (X, Z).
This motivates the construction of a family of estimates of the local variance that have the form

$$\sigma_n^2(x)^{(LA)} = \sum_{i=1}^{n} W_{n,i}(x) \cdot Z_{n,i}, \tag{3.2}$$

in dependence on the different weights. Briefly, set in this chapter $\sigma_n^2(x) := \sigma_n^2(x)^{(LA)}$.
In order to separate the influences of the random observations on the estimates of m and σ^2, we use firstly a variant defined via splitting the sample. Let the sample $D_n = \{(X_1, Y_1), \ldots,$
$(X_n, Y_n)\}$ be split into two parts

$$D_{n'} = \{(X_1, Y_1), \ldots, (X_{n'}, Y_{n'})\}$$

and

$$D_{n''} = \{(X_{n'+1}, Y_{n'+1}), \ldots, (X_n, Y_n)\},$$

where $n' = n'(n) \to \infty$, $n'' = n''(n) \to \infty$ $(n \to \infty)$ and $n = n' + n'' \geq 2$.
We use weights

$$W_{n',i}^{(1)}(x) := W_{n',i}^{(1)}(x, X_1, \ldots, X_{n'}) \quad (i = 1, \ldots, n')$$

and

$$W_{n'',i}^{(2)}(x) := W_{n'',i}^{(2)}(x, X_{n'+1}, \dots, X_{n'+n''})$$
$$(i = n'+1, \dots, n'+n'' = n).$$

Set now

$$\begin{aligned} m_{n''}^*(x) &= m_{n''}^*(x, X_{n'+1}, Y_{n'+1}, \dots, X_{n'+n''}, Y_{n'+n''}) \\ &:= \sum_{i=1}^{n''} W_{n'',i}^{(2)}(x, X_{n'+1}, \dots, X_{n'+n''}) Y_i, \end{aligned}$$

$$Z_{n,i}^* := Y_i^2 - m_{n''}^*(X_i)^2,$$

and finally

$$\sigma_n^{2*}(x) := \sum_{i=1}^{n'} W_{n',i}^{(1)}(x, X_1, \dots, X_{n'}) Z_{n,i}^*. \tag{3.3}$$

The following theorem concerns weak consistency for the estimator (3.3). Its statement is inspired by Stone's (1977) theorem on weak universal consistency of local averaging regression estimates (see also [11], Theorem 4.1).

Theorem 3.1. *Assume that the following conditions are satisfied for any distribution of X and $l = 1, 2$:*

(i) There is a constant c such that for every nonnegative measurable function f satisfying $\boldsymbol{E} f(X) < \infty$ and any n,

$$\boldsymbol{E}\left\{\sum_{i=1}^{n} |W_{n,i}^{(l)}(X, X_1, \dots, X_n)| f(X_i)\right\} \le c\boldsymbol{E} f(X).$$

(ii) There is a $D \ge 1$ such that

$$\boldsymbol{P}\left\{\sum_{i=1}^{n} |W_{n,i}^{(l)}(X, X_1, \dots, X_n)| \le D\right\} = 1.$$

for all n.

(iii) For all $a > 0$,

$$\lim_{n\to\infty} \boldsymbol{E}\left\{|W_{n,i}^{(l)}(X, X_1, \dots, X_n)| 1_{\{\|X_i - X\| > a\}}\right\} = 0.$$

(iv)

$$\sum_{i=1}^{n} W_{n,i}^{(l)}(X, X_1, \ldots, X_n) \to 1.$$

in probability.

(v)

$$\lim_{n\to\infty} \boldsymbol{E}\left\{\sum_{i=1}^{n} W_{n,i}^{2\ (l)}(X, X_1, \ldots, X_n)\right\} = 0.$$

Then the local variance function estimate σ_n^{2*} is weakly consistent for $|Y| \leq L$, i.e.,

$$\lim_{n\to\infty} \boldsymbol{E}\left\{\int \left(\sigma_n^{2*}(x) - \sigma^2(x)\right)^2 \mu(dx)\right\} = 0$$

for all distributions of (X, Y), with bounded Y.

Proof. Because of $(a+b)^2 \leq 2a^2 + 2b^2$ we have that

$$\begin{aligned}
&\int \left(\sigma_n^{2*}(x) - \sigma^2(x)\right)^2 \mu(dx) \\
&= \int \left(\sigma_n^{2*}(x) - \sum_{i=1}^{n'} W_{n',i}^{(1)}(x) Z_i + \sum_{i=1}^{n'} W_{n',i}^{(1)}(x) Z_i - \sigma^2(x)\right)^2 \mu(dx) \\
&\leq 2 \int \left(\sum_{i=1}^{n'} W_{n',i}^{(1)}(x) Z_i - \sigma^2(x)\right)^2 \mu(dx) \\
&\quad +2 \int \left[\sum_{i=1}^{n'} W_{n',i}^{(1)}(x) \left(Y_i^2 - m_{n''}^{2*}(X_i)\right)\right. \\
&\quad \left. -\sum_{i=1}^{n'} W_{n',i}^{(1)}(x) \left(Y_i^2 - m^2(X_i)\right)\right]^2 \mu(dx) \\
&= 2A_n + 2B_n.
\end{aligned}$$

Concerning A_n we have that $\boldsymbol{E}A_n \to 0$ by Stone's Theorem (see [11], Theorem 4.1), with Y_i and $m(x)$ appearing there replaced by $Y_i^2 - m^2(X_i)$ and by $\sigma^2(x)$, respectively.

Now, concerning B_n, by the Cauchy-Schwarz inequality, (ii) and boundedness of Y

$$B_n = \int \left[\sum_{i=1}^{n'} W_{n',i}^{(1)}(x) \left(m_{n''}^{2*}(X_i) - m^2(X_i) \right) \right]^2 \mu(dx)$$

$$\le \int \left[\sum_{i=1}^{n'} \sqrt{|W_{n',i}^{(1)}(x)|} \sqrt{|W_{n',i}^{(1)}(x)| |m_{n''}^{2*}(X_i) - m^2(X_i)|} \right]^2 \mu(dx)$$

$$\le \int \left(\sum_{i=1}^{n'} |W_{n',i}^{(1)}(x)| \right)$$

$$\left(\sum_{i=1}^{n'} |W_{n',i}^{(1)}(x)| \left(m_{n''}^{*}(X_i) + m(X_i) \right)^2 \left(m_{n''}^{*}(X_i) - m(X_i) \right)^2 \right) \mu(dx)$$

$$\le 2D(D^2+1)L^2 \int \sum_{i=1}^{n'} |W_{n',i}^{(1)}(x)| \left(m_{n''}^{*}(X_i) - m(X_i) \right)^2 \mu(dx). \qquad (3.4)$$

Now, working conditionally on $(X_{n'+1}, Y_{n'+1}), \ldots, (X_n, Y_n)$ we get for the expectation of the integral in (3.4)

$$\int \int \boldsymbol{E} \left[\sum_{i=1}^{n'} |W_{n',i}^{(1)}(x)| \left(m_{n''}^{*}(X_i) - m(X_i) \right)^2 \right|$$

$$\left. X_{n'+1} = x_{n'+1}, Y_{n'+1} = y_{n'+1}, \ldots, X_n = x_n, Y_n = y_n \right]$$

$$dP_{(X_{n'+1}, Y_{n'+1}, \ldots, X_n, Y_n)}(x_{n'+1}, y_{n'+1}, \ldots, x_n, y_n) \mu(dx)$$

$$= \int \int \boldsymbol{E} \left[\sum_{i=1}^{n'} |W_{n',i}^{(1)}(x)| \left(m_{n''}^{*}(X_i, x_{n'+1}, y_{n'+1}, \ldots, x_n, y_n) - m(X_i) \right)^2 \right]$$

$$dP_{(X_{n'+1}, Y_{n'+1}, \ldots, X_n, Y_n)}(x_{n'+1}, y_{n'+1}, \ldots, x_n, y_n) \mu(dx), \qquad (3.5)$$

where the last step is due to the independence of $((X_1, Y_1), \ldots, (X_{n'}, Y_{n'}))$ and $((X_{n'+1}, Y_{n'+1}), \ldots, (X_n, Y_n))$. Then, by (i) for (3.5) we get an upper bound

$$c \int \boldsymbol{E} \left\{ \left(m_{n''}^{*}(X, x_{n'+1}, y_{n'+1}, \ldots, x_n, y_n) - m(X) \right)^2 \right\} \cdot$$

$$\cdot dP_{(X_{n'+1}, Y_{n'+1}, \ldots, X_n, Y_n)}(x_{n'+1}, y_{n'+1}, \ldots, x_n, y_n)$$

$$= c\boldsymbol{E}\{ (m_{n''}^{*}(X, X_{n'+1}, Y_{n'+1}, \ldots, X_{n'+n''}, Y_{n'+n''})$$

$$- m(X))^2 \},$$

which converges to 0, again by Stone's theorem. Therefore $\boldsymbol{E}B_n \to 0$. □

Remark 3.1. Theorem 3.1 remains valid if the boundedness condition $|Y| \leq L$ a.s. is weakened to $\boldsymbol{E}\{|Y|^4\} < \infty$. This means weak universal consistency. The proof is analogous. In the final step, with abbreviation

$$m^*_{n''}(X) = m^*_{n''}(X, X_{n'+1}, Y_{n'+1}, \ldots, X_{n'+n''}, Y_{n'+n''}),$$

one has to show

$$\boldsymbol{E}\left\{(m^*_{n''}(X) + m(X))^2 (m^*_{n''}(X) - m(X))^2\right\} \to 0.$$

By the Cauchy-Schwarz inequality it suffices to show

$$\boldsymbol{E}\left\{(m^*_{n''}(X) + m(X))^4\right\} = O(1).$$

and

$$\boldsymbol{E}\left\{(m^*_{n''}(X) - m(X))^4\right\} \to 0. \tag{3.6}$$

Because of $\boldsymbol{E}\{m(X)\}^4 < \infty$ and $\boldsymbol{E}\{m^*_{n''}(X)\}^4 < \infty$ (by $\boldsymbol{E}Y^4 < \infty$), it is enough to show (3.6). According to [10], Theorem 1 with proof (compare also [11], Lemma 23.3, which deals with strong consistency), (3.6) holds under the assumptions that a finite constant c^* exists such that

$$\boldsymbol{E}\left\{\sum_{i=n'+1}^{n'+n''} \left|W^{(2)}_{n'',i}(X, X_{n'+1}, \ldots, X_{n'+n''})\right| |Y_i|\right\} \leq c^* \boldsymbol{E}\{|Y|\},$$

for integrable Y and for all n, and that (3.6) is valid in the case of bounded Y. The first assumption holds because

$$\begin{aligned}
&\boldsymbol{E}\left\{\sum_{i=n'+1}^{n'+n''} \left|W^{(2)}_{n'',i}(X, X_{n'+1}, \ldots, X_{n'+n''})\right| |Y_i|\right\} \\
&= \sum_{i=n'+1}^{n'+n''} \int \boldsymbol{E}\left\{\boldsymbol{E}\left\{|W^{(2)}_{n'',i}(x, X_{n'+1}, \ldots, X_{n'+n''})||Y_i| \Big| X_i\right\}\right\} \mu(dx) \\
&= \int \boldsymbol{E}\left\{\sum_{i=n'+1}^{n'+n''} |W^{(2)}_{n'',i}(x, X_{n'+1}, \ldots, X_{n'+n''})| \boldsymbol{E}\left\{|Y_i| \Big| X_i\right\}\right\} \mu(dx) \\
&= \boldsymbol{E}\left\{\sum_{i=n'+1}^{n'+n''} |W^{(2)}_{n'',i}(X, X_{n'+1}, \ldots, X_{n'+n''})| \boldsymbol{E}\left\{|Y_i| \Big| X_i\right\}\right\}
\end{aligned}$$

$$\leq c\boldsymbol{E}\left\{\boldsymbol{E}\left\{|Y|\middle|X\right\}\right\}$$
$$= c\boldsymbol{E}\{|Y|\}$$

by assumption (i) in Theorem 3.1. The second assumption holds, because

$$\boldsymbol{E}\left\{(m^*_{n''}(X) - m(X))^4\right\} \leq 2(D^2+1)L^2\boldsymbol{E}\left\{(m^*_{n''}(X) - m(X))^2\right\} \to 0.$$

(with bound L for $|Y|$) by Stone's theorem.

3.2 Local Variance Estimation without Splitting the Sample

In this section we give consistency of the estimator (3.2),

$$\sigma_n^2(x) = \sum_{i=1}^{n} W_{n,i}(x) \cdot Z_{n,i},$$

avoiding splitting the sample. The weights $W_{n,i}(x)$ can take different forms. An example of such weights are the partitioning weights. Introduce for that a partition $\mathcal{P}_n = \{A_{n,1}, A_{n,2}, \dots\}$ of $\mathbb{R}^d$ consisting of Borel sets $A_{n,j} \in \mathbb{R}^d$. Then

$$W_{n,i}(x, X_1, \dots, X_n) = \frac{1_{A_n(x)}(X_i)}{\sum_{l=1}^{n} 1_{A_n(x)}(X_l)} \tag{3.7}$$

are partitioning weights, where $A_n(x)$ denotes the sets $A_{n,j}$ containing $x \in \mathbb{R}^d$, using the convention $0/0 := 0$.
Further kernel weights are known in the literature, depending on the kernel $K : \mathbb{R}^d \to [0, \infty)$.

$$W_{n,i}(x, X_1, \dots, X_n) = \frac{K\left(\frac{x-X_i}{h_n}\right)}{\sum_{l=1}^{n} K\left(\frac{x-X_l}{h_n}\right)}. \tag{3.8}$$

There, let $h_n > 0$ be the bandwidth and $0/0 := 0$ again. Common kernels are for example the naive kernel ($K(x) = 1_{\{\|x\|<1\}}$) and the Epanechnikov kernel ($K(x) = (1 - \|x\|^2)_+$). For our aims we introduce now the so called boxed kernel. Define as $S_{0,r}$ the balls of Radius r and $S_{0,R}$ the balls of radius R, both centered in 0, $0 < r \leq R$. Let b be a positive constant. The symmetric boxed kernel fulfills the properties

$$1_{\{x \in S_{0,R}\}} \geq K(x) \geq b 1_{\{x \in S_{0,r}\}}.$$

with bandwidth $h_n > 0$ and $0/0 := 0$ again.

Finally, nearest neighbor weights are also frequently used, defined by

$$W_{n,i}(x, X_1, \ldots, X_n) = \frac{1}{k_n} 1_{\{X_i \text{ is among the } k_n \text{ NNs of } x \text{ in } \{X_1, \ldots, X_n\}\}} \quad (3.9)$$

($2 \leq k_n \leq n$), here usually assuming that ties occur with probability 0. This can be obtained for example via tie-breaking by indices (compare [11], pp. 86, 87).
In this case we have "large" weights in the case we find "a lot" of neighbors of X among $\{X_1, \ldots, X_n\}$. We distinguish local averaging methods in the auxiliary estimates m_n (3.1) and in the estimates σ_n^2 in (3.2), indicating the weights by $W_{n,i}^{(2)}$ and $W_{n,i}^{(1)}$ (instead of $W_{n,i}$ in (3.2)), respectively. Thus

$$m_n(X_i) = \sum_{j=1}^{n} W_{n,j}^{(2)}(X_i, X_1, \ldots, X_n) Y_j$$

where

$$W_{n,j}^{(2)}(x, X_1, \ldots, X_n),$$

is of partitioning type, with partitioning sequence $\left\{A_{n,j}^{(2)}\right\}$, or of kernel type, with kernel $K^{(2)}$ and bandwidths $h_n^{(2)}$, or of nearest neighbor type, with $k_n^{(2)}$ neighbors.
Now with $Z_{n,i} = Y_i^2 - m_n^2(X_i)$ we define a family of estimators of the local variance function by

$$\sigma_n^2(x) = \sum_{i=1}^{n} W_{n,i}^{(1)}(x) \cdot Z_{n,i}, \quad (3.10)$$

depending on weights

$$W_{n,i}^{(1)}(x) = W_{n,i}^{(1)}(x, X_1, \ldots, X_n),$$

that are of partitioning type, with partitioning sequence $\left\{A_{n,j}^{(1)}\right\}$, or of kernel type, with kernel $K^{(1)}$ and bandwidths $h_n^{(1)}$. Nearest neighbor weights will not be used for $W_{n,i}^{(1)}(x)$, because for these we would need other arguments in the proof of consistency than for partitioning and kernel weights.

The following theorem states the consistency of the estimator (3.10) depending on the weights defined in (3.7), (3.8) and (3.9).

Theorem 3.2. *Let (X,Y) have an arbitrary distribution with $\boldsymbol{E}\{Y^4\} < \infty$. For partitioning weights defined according to (3.7) assume that, for each sphere S centered at the origin*

$$\lim_{n\to\infty} \max_{j:\ A_{n,j}^{(l)} \cap S \neq \emptyset} diam(A_{n,j}^{(l)}) = 0, \quad l = 1, 2, \tag{3.11}$$

$$\lim_{n\to\infty} \frac{|\{j : A_{n,j}^{(l)} \cap S \neq \emptyset\}|}{n} = 0, \quad l = 1, 2. \tag{3.12}$$

For kernel weights defined according to (3.8) with kernels $K^{(l)}$ assume that the bandwidths satisfy

$$0 < h_n^{(l)} \to 0, \quad nh_n^{(l)d} \to \infty, \quad l = 1, 2, \tag{3.13}$$

($K^{(l)}$ symmetric, $1_{S_{0,R}}(x) \geq K^{(l)}(x) \geq b 1_{S_{0,r}}(x)$ $(0 < r \leq R < \infty,\ b > 0)$). For nearest neighbor weights defined according to (3.9) assume that

$$2 \leq k_n^{(2)} \leq n, \quad k_n^{(2)} \to \infty, \quad \frac{k_n^{(2)}}{n} \to 0 \tag{3.14}$$

Then for the estimate (3.10) under the above assumptions

$$\lim_{n\to\infty} \boldsymbol{E} \int \left|\sigma_n^2(x) - \sigma^2(x)\right| \mu(dx) = 0$$

holds.
(Universal consistency of the local averaging estimator of the local variance)

Theorem 3.2 will be proven by Lemmas 3.1 and 3.3.
The following Lemma 3.1 modifies Remark 5 in Kohler [16]. It is within the framework that the dependent variable Y can be observed only with supplementary, maybe correlated, measurement errors. Since it is not assumed that the means of these measurement errors are zero, these kinds of errors are not already included in standard models. Therefore, the dataset is of the form

$$\overline{D}_n = \{(X_1, \overline{Y}_{1,n}), \ldots, (X_n, \overline{Y}_{n,n})\},$$

where

$$\overline{Y}_i = m(X_i) + \overline{\epsilon}_i \quad \text{admits} \quad \boldsymbol{E}\{\overline{\epsilon}_i | X_i\} \neq 0,$$

in contrast to the common regression model where $\boldsymbol{E}\{\epsilon_i | X_i\} = 0$.

Lemma 3.1. *Let $\overline{m}_n$ be local averaging estimators of m of the form*

$$\overline{m}_n(x) = \sum_{i=1}^{n} W_{n,i}(x)\overline{Y}_i$$

with $\overline{Y}_i = \overline{Y}_{n,i}$. Assume that the weights $W_{n,i}(x) = W_{n,i}(x, X_1, \ldots, X_n)$ are of partitioning type (3.7) or kernel type (3.8) with $1_{S_{0,R}}(x) \geq K(x) \geq b1_{S_{0,R}}$ for some $0 < R < \infty$, $b > 0$, satisfying (3.11) $\wedge$ (3.12) and (3.13), respectively. Further assume

$$\mathbf{E}\left\{Y^2\right\} < \infty, \quad \mathbf{E}\left\{\overline{Y}_i^2\right\} < \infty \quad (i = 1, \ldots, n)$$

and

$$\mathbf{E}\left\{\frac{1}{n}\sum_{i=1}^{n}|\overline{Y}_i - Y_i|^2\right\} \to 0. \tag{3.15}$$

Then

$$\mathbf{E}\left\{\int |\overline{m}_n(x) - m(x)|\mu(dx)\right\} \to 0.$$

Proof. As Kohler ([16], Remark 5) suggested, further by use of the Cauchy-Schwarz inequality,

$$\begin{aligned}
&\mathbf{E}\left\{\int\left|\sum_{i=1}^{n} W_{n,i}(x)\overline{Y}_i - m(x)\right|\mu(dx)\right\}\\
&= \mathbf{E}\left\{\int\left|\sum_{i=1}^{n} W_{n,i}(x)\left[\overline{Y}_i - Y_i + Y_i\right] - m(x)\right|\mu(dx)\right\}\\
&\leq \mathbf{E}\left\{\int\left|\sum_{i=1}^{n} W_{n,i}(x)Y_i - m(x)\right|\mu(dx)\right\}\\
&\quad + \mathbf{E}\left\{\int\left|\sum_{i=1}^{n} W_{n,i}(x)\left[\overline{Y}_i - Y_i\right]\right|\mu(dx)\right\}\\
&\leq \mathbf{E}\left(\left\{\int\left|\sum_{i=1}^{n} W_{n,i}(x)Y_i - m(x)\right|^2\mu(dx)\right\}\right)^{\frac{1}{2}}\\
&\quad + \mathbf{E}\left\{\int\sum_{i=1}^{n} W_{n,i}(x)|\overline{Y}_i - Y_i|\mu(dx)\right\}\\
&= K_{n,1} + K_{n,2}.
\end{aligned}$$

$K_{n,1}^2$ is simply the expected L_2-error of the local averaging estimate of m on the basis of observed $((X_1, Y_1), \ldots, (X_n, Y_n))$. By Theorem 4.2 and Theorem 5.1 in [11], respectively, $K_{n,1} \to 0$. It remains to show $K_{n,2} \to 0$. We consider only the kernel case. The partitioning case can be treated analogously. It holds, where $S_{x,R}$ is the sphere with radius R centered in x

$$K_{n,2} \leq \frac{1}{b}\boldsymbol{E}\left\{\int \frac{\sum_{i=1}^n |\overline{Y}_i - Y_i| 1_{S_{0,Rh_n}}(x - X_i)}{\sum_{i=1}^n 1_{S_{0,Rh_n}}(x - X_i)}\mu(dx)\right\}$$

$$= \frac{1}{b}\int \boldsymbol{E}\left\{\sum_{i=1}^n \frac{|\overline{Y}_i - Y_i| 1_{S_{0,Rh_n}}(x - X_i)}{1 + \sum_{j\in\{1,\ldots,n\}\setminus\{i\}} 1_{S_{0,Rh_n}}(x - X_j)}\right\}\mu(dx)$$

$$= \frac{1}{b}\sum_{i=1}^n \int \boldsymbol{E}\left\{\boldsymbol{E}\left\{\frac{|\overline{Y}_i - Y_i| 1_{S_{0,Rh_n}}(x - X_i)}{1 + \sum_{j\in\{1,\ldots,n\}\setminus\{i\}} 1_{S_{x,Rh_n}}(X_j)}\middle| X_i\right\}\right\}\mu(dx)$$

$$\leq \frac{1}{b}\sum_{i=1}^n \int \boldsymbol{E}\left\{1_{S_{0,Rh_n}}(x - X_i)\left(\boldsymbol{E}\left\{|\overline{Y}_i - Y_i|^2 | X_i\right\}\right)^{1/2}\right\}$$

$$\left(\boldsymbol{E}\left\{\frac{1}{(1 + \sum_{j\in\{1,\ldots,n\}\setminus\{i\}} 1_{S_{x,Rh_n}}(X_j))^2}\right\}\right)^{1/2}\mu(dx)$$

(by the Cauchy-Schwarz inequality for conditional expectations and independence of the pair $(X_i, (X_1, \ldots, X_{i-1}, X_{i+1}, \ldots, X_n))$)

$$\leq \frac{1}{b}\sum_{i=1}^n \int \boldsymbol{E}\left\{1_{S_{0,Rh_n}}(x - X_i)\left(\boldsymbol{E}\left\{|\overline{Y}_i - Y_i|^2 | X_i\right\}\right)^{1/2}\right\}\frac{1}{n\mu(S_{x,Rh_n})}\mu(dx)$$

(because the sum in the denominator is $b(n-1, \mu(S_{x,Rh_n}))$-distributed, compare [11], Lemma 4.1)

$$\leq \frac{c^*}{n}\sum_{i=1}^n \boldsymbol{E}\left\{\left(\boldsymbol{E}\left\{|\overline{Y}_i - Y_i|^2 | X_i\right\}\right)^{1/2}\right\}$$

(for some constant c^*, by the covering lemma ([11], Lemma 23.6))

$$\leq \frac{c^*}{n}\sum_{i=1}^n \left(\boldsymbol{E}\left\{\boldsymbol{E}\left\{|\overline{Y}_i - Y_i|^2 | X_i\right\}\right\}\right)^{1/2}$$

(by the Cauchy-Schwarz inequality)

$$= \frac{c^*}{n} \sum_{i=1}^{n} \left(\boldsymbol{E}\left\{ |\overline{Y}_i - Y_i|^2 \right\} \right)^{1/2}$$

$$\leq c^* \left(\frac{1}{n} \sum_{i=1}^{n} \boldsymbol{E}\left\{ |\overline{Y}_i - Y_i|^2 \right\} \right)^{1/2}$$

(once more by the Cauchy-Schwarz inequality).
Therefore $K_{n,2} \to 0$ by (3.15). □

Lemma 3.2. *Under the assumptions of Lemma 3.1 but*

$$\frac{1}{n} \sum_{i=1}^{n} |\overline{Y}_i - Y_i|^2 \xrightarrow{P} 0 \quad (n \to \infty)$$

instead of (3.15), one has

$$\int |\overline{m}_n(x) - m(x)| \mu(dx) \xrightarrow{P} 0. \tag{3.16}$$

Proof. Let $r \in \mathbb{N}$ be arbitrary. Let (n_k) be an arbitrary subsequence of indices. Then a subsequence (n_{k_l}) of (n_k) exists such that

$$\frac{1}{n_{k_l}} \sum_{i=1}^{n_{k_l}} |\overline{Y}_{n_{k_l},i} - Y_i|^2 \to 0 \quad a.s.$$

By Egorov's theorem a set $\Omega_r((n_k))$ with $\boldsymbol{P}(\Omega_r^c) \leq \frac{1}{r}$ exists such that

$$\frac{1}{n_{k_l}} \sum_{i=1}^{n_{k_l}} |\overline{Y}_{n_{k_l},i} - Y_i|^2 \to 0 \quad \text{uniformly on } \Omega_r,$$

with square integrability of each $\overline{Y}_{n_{k_l},i}$ on Ω_r. We repeat the proof of Lemma 3.1, replacing the index n by n_{k_l} and providing the random variables with the factor 1_{Ω_δ}. Especially we obtain

$$\int \boldsymbol{E}\left\{ \sum_{i=1}^{n_{k_l}} W_{n_{k_l},i}(x) \boldsymbol{E}\left\{ |\overline{Y}_{n_{k_l},i} - Y_i| 1_{\Omega_r} | X_i \right\} \right\} \mu(dx)$$

$$\leq \frac{c}{(n_{k_l})^{1/2}} \left(\boldsymbol{E}\left\{ \sum_{i=1}^{n_{k_l}} |\overline{Y}_{n_{k_l},i} - Y_i| 1_{\Omega_r} \right\} \right)^{1/2} \to 0$$

and summarizing

$$\mathbf{E}\left\{\int |\overline{m}_{n_{k_l}}(x) - m(x)|1_{\Omega_r}\mu(dx)\right\} \to 0$$

therefore

$$\int |\overline{m}_{n_{k_{l_j}}}(x) - m(x)|\mu(dx)1_{\Omega_r} \xrightarrow{P} 0,$$

and thus

$$\int |\overline{m}_{n_{k_{l_j}}}(x) - m(x)|\mu(dx)1_{\Omega_r} \to 0 \quad a.s.$$

for a suitable sequence $(n_{k_{l_j}})$ of (n_{k_l}). This means that for each $r \in \mathbb{N}$ and each subsequence (n') of indices a subsequence (n'') exists with

$$\int |\overline{m}_{n''}(x) - m(x)|\mu(dx)1_{\Omega_r}((n')) \to 0 \quad a.s.$$

By Cantor's diagonal method and because of $\mathbf{P}(\cap\ \Omega_r^c) = 0$ we obtain that for each subsequence (n^*) of indices a subsequence (n^{**}) exists with

$$\int |\overline{m}_{n^{**}}(x) - m(x)|^2\mu(dx) \to 0 \quad a.s.$$

Thus the assertion (3.16) is obtained. □

Lemma 3.3. *Let m_n be local averaging estimators of m of the form*

$$m_n(x) = \sum_{i=1}^{n} W_{n,i}(x)Y_i.$$

Assume that the weights $W_{n,i}(x) = W_{n,i}(x, X_1, \ldots, X_n)$ are of partitioning type (3.7) or of kernel type (3.8) with $1_{S_{0,R}}(x) \geq K(x) \geq b1_{S_{0,r}}(x)$ for some $0 < r \leq R < \infty$, $b > 0$, or of nearest neighbor type (3.9) (here under assumption that ties occur with probability 0), satisfying (3.11) $\wedge$ (3.12), (3.13) and (3.14), respectively. Further assume $\mathbf{E}\{Y^2\} < \infty$. Then

$$\mathbf{E}\{|m_n(X_1) - m(X_1)|^2\} \to 0.$$

If moreover $\mathbf{E}\{Y^4\} < \infty$, then

$$\mathbf{E}\{|m_n(X_1) - m(X_1)|^4\} \to 0.$$

Proof. We first assume that $\mathbf{E}\{Y^2\} < \infty$ and that $W_{n,j}$ is of kernel type. Then

$$\begin{aligned}
&\mathbf{E}\{|m_n(X_1) - m(X_1)|^2\}\\
&= \mathbf{E}\left\{\left|\frac{Y_1K(0) + \sum_{j=2}^n Y_jK\left(\frac{X_1-X_j}{h_n}\right)}{K(0) + \sum_{j=2}^n K\left(\frac{X_1-X_j}{h_n}\right)} - m(X_1)\right|^2\right\}\\
&\leq 2K(0)^2\mathbf{E}\left\{\frac{\mathbf{E}\left\{Y^2|X\right\}}{\left(K(0) + \sum_{j=2}^n K\left(\frac{X-X_j}{h_n}\right)\right)^2}\right\}\\
&+ 2\mathbf{E}\left\{\left|\frac{\sum_{j=2}^n Y_jK\left(\frac{X-X_j}{h_n}\right)}{K(0) + \sum_{j=2}^n K\left(\frac{X-X_j}{h_n}\right)} - m(X)\right|^2\right\}.
\end{aligned}$$

The second term of the right-hand side converges to 0 as in the proof of Theorem 5.1 in [11]. With a suitable finite constant c_1 the first term is bounded by

$$\begin{aligned}
&c_1\int \mathbf{E}\left\{Y^2\middle|X = x\right\}\mathbf{E}\left\{\frac{1}{\left(1 + \sum_{j=2}^n 1_{S_{x,rh_n}}(X_j)\right)^2}\right\}\mu(dx)\\
&\leq c_1\int \mathbf{E}\left\{Y^2|X = x\right\}\frac{1}{n\mu(S_{x,rh_n})}\mu(dx) \to 0
\end{aligned}$$

by Lemma 4.1 (i) in [11] together with Lemma 24.6 in [11], $nh_n^d \to \infty$ in assumption (3.13), $\mathbf{E}\{Y^2\} < \infty$ and the dominated convergence theorem. The case that $W_{n,j}$ is defined via partitioning, is treated analogously by use of Theorem 4.2 in [11], and for $\epsilon > 0$

$$\begin{aligned}
&\int \mathbf{E}\left\{Y^2|X = x\right\}\mathbf{E}\frac{1}{\left(1 + \sum_{j=2}^n 1_{A_n(x)}(X_j)\right)^2}\mu(dx)\\
&\leq \int \mathbf{E}\{Y^2|X = x\}1_{\{\mathbf{E}\{Y^2|X=x\}>L\}}\mu(dx)\\
&\quad + \int \mathbf{E}\{Y^2|X = x\}1_{\{\mathbf{E}\{Y^2|X=x\}\leq L\}}\mathbf{E}\left\{\frac{1}{1 + \sum_{j=2}^n 1_{A_n(x)}(X_j)}\right\}\mu(dx)\\
&\leq \epsilon + L\int_{S^c}\mu(dx) + L\int_S \frac{1}{n\mu(A_n(x))}\mu(dx)\\
&\quad \text{(for suitable L} < \infty \text{ and by Lemma 4.1 (i) in [11])}\\
&\leq 2\epsilon + L\int_S \frac{1}{n\mu(A_n(x))}\mu(dx)
\end{aligned}$$

(for suitable sphere S centered at 0)

$= 2\epsilon + o(1)$

(by assumption (3.12)).

In the case that $W_{n,j}$ is defined via nearest neighbors, we write

$$\begin{aligned}
&\boldsymbol{E}\{|m_n(X_1) - m(X_1)|^2\} \\
&= \boldsymbol{E}\left\{\left|\frac{Y_1 + D_n}{1 + (k_n - 1)} - m(X_1)\right|^2\right\}
\end{aligned}$$

with

$$D_n = \sum_{j=2}^{n} Y_j 1_{\{X_j \text{ is among the } (k_n-1) \text{ NNs of } X_1 \text{ in } \{X_2,\ldots,X_n\}\}},$$

and notice

$$\boldsymbol{E}\left\{\left|\frac{D_n}{k_n - 1} - m(X_1)\right|^2\right\} \to 0$$

by Theorem 6.1 in [11]. Further

$$\boldsymbol{E}\left\{\left|\frac{Y_1 + D_n}{k_n} - \frac{D_n}{k_n - 1}\right|^2\right\} \leq \frac{2}{k_n^2}\left(\boldsymbol{E}\left\{Y_1^2\right\} + \boldsymbol{E}\left\{\left(\frac{D_n}{k_n - 1}\right)^2\right\}\right) \to 0$$

because of

$$\boldsymbol{E}\left\{Y_1^2\right\} < \infty, \quad \boldsymbol{E}\left\{\left(\frac{D_n}{k_n - 1}\right)^2\right\} \to \boldsymbol{E}\{m(X_1)^2\} < \infty, k_n \to \infty.$$

Now, we consider the case $\boldsymbol{E}\{Y^4\} < \infty$. The above proof shows that for $r = 2, 4$ one has the representation

$$\begin{aligned}
J_n^{(r)} &:= \boldsymbol{E}\{|m_n(X_1) - m(X_1)|^r\} \\
&= \boldsymbol{E}\left\{\left|\sum_{i=1}^{n} Y_i \overline{W}_{n,i}(X, X_2, \ldots, X_n) - m(X)\right|^r\right\}
\end{aligned}$$

for some $\overline{W}_{n,i} \geq 0$ with $\sum_{i=1}^{n} \overline{W}_{n,i} = 1$, e.g., in the kernel case $\overline{W}_{n,i}$ with $K(0)$ instead of $K\left(\frac{\cdot - X_1}{h_n}\right)$ in $W_{n,i}$. Then by Györfi [10], Theorem 1 with proof (compare also [11], Lemma 23.3 with proof, and [34], last part of Lemma 8 with $\delta = 1$, $p = 2$ and convergence in probability instead of almost sure convergence), $J_n^{(2)} \to 0$ for $\boldsymbol{E}\{Y^2\} < \infty$ (already proven) implies $J_n^{(4)} \to 0$ for $\boldsymbol{E}\{Y^4\} < \infty$. □

Proof of Theorem 3.2. We apply Lemma 3.1 with Y_i, $\overline{Y}_i$, $W_{n,i}$, $\overline{m}_n$ and m replaced by $Y_i^2 - m^2(X_i)$, $Y_i^2 - m_n^2(X_i)$, $W_{n,i}^{(1)}$, σ_n^2 and σ^2 in Theorem 3.2, respectively. We notice

$$\begin{aligned}
&\boldsymbol{E}\left\{\frac{1}{n}\sum_{i=1}^{n}|(Y_i^2 - m_n^2(X_i)) - (Y_i^2 - m^2(X_i))|^2\right\} \\
&= \boldsymbol{E}\{|m_n^2(X_1) - m^2(X_1)|^2\} \\
&\quad \text{(due to symmetry with respect to } (X_1,Y_1),\dots,(X_n,Y_n)) \\
&\le \left(\boldsymbol{E}\left\{|m_n(X_1) + m(X_1)|^4\right\}\right)^{\frac{1}{2}} \left(\boldsymbol{E}\left\{|m_n(X_1) - m(X_1)|^4\right\}\right)^{\frac{1}{2}} \\
&\quad \text{(because of the Cauchy-Schwarz inequality)} \\
&\to 0.
\end{aligned}$$

The latter is obtained by the triangle inequality

$$\begin{aligned}
&\left(\boldsymbol{E}\left\{|m_n(X_1) + m(X_1)|^4\right\}\right)^{\frac{1}{4}} \\
&\le \left(\boldsymbol{E}\left\{|m_n(X_1) - m(X_1)|^4\right\}\right)^{\frac{1}{4}} + 2\boldsymbol{E}(\{|m(X_1)|^4\})^{\frac{1}{4}},
\end{aligned}$$

$$\begin{aligned}
\boldsymbol{E}\{m(X_1)^4\} &= \boldsymbol{E}\{m(X)^4\} \\
&= \boldsymbol{E}\{(\boldsymbol{E}(Y|X))^4\} \le \boldsymbol{E}\{\boldsymbol{E}\{Y^4|X\}\} = \boldsymbol{E}\{Y^4\} < \infty
\end{aligned}$$

because of Jensen's inequality for conditional expectations,

$$\boldsymbol{E}\left\{|m_n(X_1) - m(X_1)|^4\right\} \to 0$$

because of Lemma 3.3 with $W_{n,i}$ there replaced by $W_{n,i}^{(2)}$ in Theorem 3.2. Thus Lemma 3.1 yields the assertion.

3.3 Rate of Convergence

In this section we establish a rate of convergence for the estimate of the local variance defined in Section 3.2.

Theorem 3.3. *Let the estimate of the local variance σ^2 be given by (1.5) with weights $W_{n,i}^{(1)}(x)$ of cubic partition with side length $h_n^{(1)}$ or with naive*

kernel $1_{S_{0,1}^{(1)}}$ *with bandwidths* $h_n^{(1)}$, *further for* $m_n(X_i)$ *given by (1.4) with weights* $W_{n,i}^{(2)}(x)$ *of cubic partition with side length* $h_n^{(2)}$ *or with naive kernel* $1_{S_{0,1}^{(2)}}$ *and bandwidths* $h_n^{(2)}$ *or with* $k_n^{(2)}$*-nearest neighbors (the latter for* $d \geq 2$*).*
Assume that X *is bounded and that*

$$|Y| \leq L \in [0, \infty),$$

$$|m(x) - m(z)| \leq C\|x - z\|, \quad x, z \in \mathbb{R}^d,$$

and finally, that

$$|\sigma^2(x) - \sigma^2(z)| \leq D\|x - z\|, \quad x, z \in \mathbb{R}^d$$

($\| \ \|$ denoting the Euclidean norm). Then, for

$$h_n^{(1)} \sim n^{-\frac{1}{d+2}},$$

and

$$h_n^{(2)} \sim n^{-\frac{1}{d+2}}, \quad \text{and} \quad k_n^{(2)2} \sim n^{\frac{2}{d+2}}, \quad \text{respectively,}$$

$$\boldsymbol{E} \int |\sigma_n^2(x)^{(LA)} - \sigma^2(x)|\mu(dx) = O\left(n^{-\frac{1}{d+2}}\right).$$

(Rate of convergence of the local averaging estimator of the local variance)

Proof. As in the proof of Lemma 3.1 and by the Cauchy-Schwarz inequality we obtain

$$\begin{aligned}
&\boldsymbol{E}\left\{\int \left| \left(\sum_{i=1}^n W_{n,i}^{(1)}(x) Z_{n,i}\right) - \sigma^2(x) \right| \mu(dx)\right\} \\
&\leq \boldsymbol{E}\left\{\int \left| \sum_{i=1}^n W_{n,i}^{(1)}(x) \left(Y_i^2 - m^2(X_i)\right) - \sigma^2(x) \right| \mu(dx)\right\} \\
&\quad + \boldsymbol{E}\left\{\int \left| \sum_{i=1}^n W_{n,i}^{(1)}(x) \left[m_n^2(X_i) - m^2(X_i)\right] \right| \mu(dx)\right\} \\
&\leq \boldsymbol{E}\left\{\int \left| \sum_{i=1}^n W_{n,i}^{(1)}(x) \left(Y_i^2 - m^2(X_i)\right) - \sigma^2(x) \right| \mu(dx)\right\} \\
&\quad + c\left(\frac{1}{n}\sum_{i=1}^n \boldsymbol{E}\left\{\left|m_n^2(X_i) - m^2(X_i)\right|^2\right\}\right)^{\frac{1}{2}}
\end{aligned}$$

$$\leq \boldsymbol{E}\left\{\int\left|\sum_{i=1}^{n}W_{n,i}^{(1)}(x)\left(Y_i^2 - m^2(X_i)\right) - \sigma^2(x)\right|^2 \mu(dx)\right\}$$
$$+c^*\left(\boldsymbol{E}\left\{|m_n(X_1) - m(X_1)|^2\right\}\right)^{\frac{1}{2}}$$
$$=: K_n + c^* L_n$$

with suitable $c^* \in [0, \infty)$ because of boundedness of Y and by symmetry. We have

$$K_n$$
$$\leq \left(\boldsymbol{E}\left\{\int\left|\sum_{i=1}^{n}W_{n,i}^{(1)}(x)\left(Y_i^2 - m^2(X_i)\right) - \sigma^2(x)\right|^2 \mu(dx)\right\}\right)^{\frac{1}{2}}$$
$$= O\left(n^{-\frac{1}{d+2}}\right)$$

by the Cauchy-Schwarz inequality and by Theorems 4.3, 5.2 and 6.2 in [11], respectively.
According to the proof of Lemma 3.3 these theorems together with boundedness of Y (with sphere $S \supset S^*$, centered in 0),

$$\int_S \frac{1}{n\mu(A_n^{(2)}(x))}\mu(dx) = O\left(\frac{1}{nh_n^{(2)d}}\right) = O\left(n^{-\frac{2}{d+2}}\right)$$

(because the number of cubes in S is $O\left(\frac{1}{h_n^{(2)d}}\right)$) in the partitioning case,

$$\int_S \frac{1}{n\mu(S_{x,rh_n^{(2)}})}\mu(dx) = O\left(\frac{1}{nh_n^{(2)d}}\right) = O\left(n^{-\frac{2}{d+2}}\right)$$

(by (5.1) in [11]) in the kernel case,

$$\boldsymbol{E}\left\{\left|\frac{Y_1 + D_n}{k_n^{(2)}} - \frac{D_n}{k_n^{(2)} - 1}\right|^2\right\} = O\left(\frac{1}{k_n^{(2)2}}\right) = O\left(n^{-\frac{2}{d+2}}\right)$$

in the nearest neighbor case, yield

$$L_n^2 = O\left(n^{-\frac{2}{d+2}}\right).$$

3.4 Local Variance Estimation with Additional Measurement Errors

We recall the important variable $Z := Y^2 - m^2(X)$ and their corresponding observations in the case of known regression function $Z_i := Y_i^2 - m^2(X_i)$. In the general case m is however to be estimated. One can do that by use of least squares estimates $m_n = m^{(LS)}$ or by local averaging estimates $m_n = m_n^{(LA)}$ (as in Theorem 3.2) of partitioning type satisfying (3.11) $\wedge$ (3.12) with $l = 2$, additionally assuming that the partitioning sequence $(\{A_{n,j}^{(2)}\})$ is nested, i.e., $A_{n+1}^{(2)}(x) \subset A_n^{(2)}(x)$ for all $n \in \mathbb{N}$, $x \in \mathbb{R}^d$, or of kernel type ($K^{(2)}$ symmetric, $1_{S_{0,R}} \geq K^{(2)}(x) \geq b 1_{S_{0,r}}(x)$ $(0 < r \leq R < \infty,\ b > 0)$) with bandwidths h_n satisfying (3.13) with $l = 2$:

$$Z_{n,i} = Y_i^2 - m_n^2(X_i).$$

With additional noise these variables are taking the form

$$\overline{Z}_{n,i} = \overline{Y}_i^2 - \overline{m}_n^2(X_i).$$

with the noisy data $\overline{Y}_i$ used also in the corresponding definition of $\overline{m}_n = \overline{m}^{(LS)}$ and $\overline{m}_n = \overline{m}^{(LA)}$, respectively. We refer the reader to Kohler [16] for the topics concerning the regression estimates $m^{(LS)}$, by use of piecewise polynomials, see also [11], Chapter 19, especially Section 19.4 and Problems and Exercises.

Let a family of estimates of the local variance function in case of additional measurement errors for the dependent variable Y be given by

$$\overline{\sigma}_n^2(x) := \overline{\sigma}_n^2(x)^{(LA)} = \sum_{i=1}^{n} W_{n,i}^{(1)}(x) \cdot \overline{Z}_{n,i}, \tag{3.17}$$

with weights $W_{n,i}^{(1)}(x) = W_{n,i}^{(1)}(x, X_1, \ldots, X_n)$ of partitioning or of kernel type with kernel $K^{(1)}$ and bandwidths $h_n^{(1)}$, respectively, satisfying (3.11) $\wedge$ (3.12) and (3.13) with $l = 1$, respectively. For this estimator we show now consistency.

Theorem 3.4. *Let the assumptions of Theorem 3.2 hold and additionally let the difference between Y_i and the noisy data $\overline{Y}_i$ satisfy*

$$\frac{1}{n}\sum_{i=1}^{n} \left(Y_i - \overline{Y}_i\right)^2 \xrightarrow{P} 0. \tag{3.18}$$

For Y_i and $\overline{Y}_i$ assume uniform boundedness: $|Y| \leq L$, $|\overline{Y}_i| \leq L$ for some $L \in [0, \infty)$.
Then, for the estimate (3.17) with $\overline{m}_n = \overline{m}_n^{(LA)}$

$$\lim_{n\to\infty} \boldsymbol{E} \int \left(\overline{\sigma}_n^2(x) - \sigma^2(x)\right)^2 \mu(dx) = 0$$

holds.
(Consistency of the local averaging estimator of the local variance with additional measurements error in the response variable)

Proof. We apply Lemma 3.1 with Y_i, $\overline{Y}_i$, $\overline{m}_n$ and m replaced by $Y_i^2 - m^2(X_i)$, $\overline{Y}_i^2 - \overline{m}_n^2(X_i)$, σ_n^2 and σ^2, respectively. If we prove that

$$\boldsymbol{E}\left\{\frac{1}{n}\sum_{i=1}^{n}|(\overline{Y}_i^2 - \overline{m}_n^2(X_i)) - (Y_i^2 - m^2(X_i))|^2\right\} \to 0,$$

(analogously to (3.15)), the assertion follows.
The left-hand side is bounded by

$$2\boldsymbol{E}\left\{\frac{1}{n}\sum_{i=1}^{n}|\overline{Y}_i^2 - Y_i^2|^2 + \frac{1}{n}\sum_{i=1}^{n}|\overline{m}_n(X_i)^2 - m^2(X_i)|^2\right\}$$
$$\leq c'\left(\boldsymbol{E}\frac{1}{n}\sum_{i=1}^{n}|\overline{Y}_i - Y_i|^2 + \boldsymbol{E}\frac{1}{n}\sum_{i=1}^{n}|\overline{m}_n(X_i) - m(X_i)|^2\right),$$

for some finite constant c' because of the uniform boundedness assumption. Because of (3.18) and the uniform boundedness assumption we have

$$\boldsymbol{E}\left\{\frac{1}{n}\sum_{i=1}^{n}|\overline{Y}_i - Y_i|^2\right\} \to 0 \tag{3.19}$$

by the dominated convergence theorem. For $\overline{m}_n^{(LA)}$ we notice that by [11], Lemma 24.11 and Lemma 24.7 (Hardy-Littlewood) and its extension (24.10) together with pp. 595, 503, 504, respectively, (for the empirical measure with respect to $(X_1, \ldots, X_n)$ and the function $x_i \to \overline{y}_i - y_i$ $(i = 1, \ldots, n)$ for the realizations $(x_i, y_i, \overline{y}_i)$ of $(X_i, Y_i, \overline{Y}_i)$, without sup)

$$\frac{1}{n}\sum_{i=1}^{n}\left|\frac{\frac{1}{n}\sum_{j=1}^{n}(\overline{Y}_j - Y_j)1_{A_n(X_j)}(X_i)}{\frac{1}{n}1_{A_n(X_j)}(X_i)}\right|^2 \leq c^*\frac{1}{n}\sum_{i=1}^{n}(\overline{Y}_i - Y_i)^2$$

and

$$\frac{1}{n}\sum_{i=1}^{n}\left|\frac{\frac{1}{n}\sum_{j=1}^{n}(\overline{Y}_j - Y_j)K\left(\frac{X_i-X_j}{h_n}\right)}{\frac{1}{n}K\left(\frac{X_i-X_j}{h_n}\right)}\right|^2 \leq c^*\frac{1}{n}\sum_{i=1}^{n}(\overline{Y}_i - Y_i)^2,$$

respectively, for some finite constant c^*, thus

$$\boldsymbol{E}\left\{\frac{1}{n}\sum_{i=1}^{n}|\overline{m}_n^{(LA)}(X_i) - m_n^{(LA)}(X_i)|^2\right\} \leq c^*\boldsymbol{E}\left\{\frac{1}{n}\sum_{i=1}^{n}|\overline{Y}_i - Y_i|^2\right\} \to 0.$$

Further

$$\begin{aligned}
&\boldsymbol{E}\left\{\frac{1}{n}\sum_{i=1}^{n}|m_n^{(LA)}(X_i) - m(X_i)|^2\right\} \\
&= \boldsymbol{E}\left\{\frac{1}{n}|m_n^{(LA)}(X_1) - m(X_1)|^2\right\} \\
&\quad \text{(by symmetry)} \\
&\to 0 \text{ (by Lemma 3.3).}
\end{aligned}$$

Therefore

$$\boldsymbol{E}\left\{\frac{1}{n}\sum_{i=1}^{n}|\overline{m}_n^{(LA)}(X_i) - m(X_i)|^2\right\} \to 0 \tag{3.20}$$

□

Chapter 4
Partitioning Estimation via Nearest Neighbors

4.1 Introduction

Let Y be a square integrable real valued random variable and let X be a d-dimensional random vector, taking values in the space $\mathbb{R}^d$. The task of regression analysis is to estimate Y given X, i.e., to find a measurable function $f : \mathbb{R}^d \to \mathbb{R}$, such that $f(X)$ is a "good approximation" of Y, that is, $|f(X) - Y|$ has to be "small". The "closeness" of $f(X)$ to Y is typically measured by the so-called **mean squared error** of f,

$$\boldsymbol{E}\{(Y - f(X))^2\}.$$

It is well known that the regression function m minimizes this error (where $m(x) := \boldsymbol{E}\{Y|X = x\}$),

$$V := \min_f \boldsymbol{E}\{(Y - f(X))^2\} = \boldsymbol{E}\{(Y - m(X))^2\}.$$

V, the so-called residual variance (global variance), is a measure of how close we can get to Y using any measurable function f. It indicates how difficult a regression problem is. Since the distribution of (X, Y), and therefore m, are unknown, one is interested in estimating V by use of data observations

$$D_n = \{(X_1, Y_1), \ldots, (X_n, Y_n)\}, \tag{4.1}$$

which are independent copies of (X, Y) (random design).
A related interesting problem is the estimation of the local variance (or conditional variance), defined as

$$\sigma^2(x) := \boldsymbol{E}\{(Y - m(X))^2|X = x\} = \boldsymbol{E}\{Y^2|X = x\} - m^2(x). \tag{4.2}$$

It holds

$$V = \boldsymbol{E}\{\sigma^2(X)\}. \tag{4.3}$$

Liitiäinen, Corona and Lendasse [19], with generalization in Liitiäinen, Corona and Lendasse [20], investigated a nonparametric estimator of the residual variance V, introduced by [6, 7], which is based on first and second nearest neighbors and the differences of the corresponding response variables. They obtained mean square convergence under bounded conditional fourth moment of Y and convergence order $n^{-2/d}$ for $d \geq 2$ under finite suitable moments of X and under Lipschitz continuity of m. It simplifies an estimator given in [4] with the same convergence order, based on first nearest neighbors. Note that the plug-in method based on a consistent regression estimate only yields the convergence order $n^{-2/(d+2)}$ for $d \geq 2$ (Theorem 3.1, [4]). On the other hand, in the case of a very smooth regression function suitable regression estimates lead to better rates of convergence. [25] investigated a U-statistic-based estimator of V.

In this chapter, first we show strong consistency of the (global) residual variance estimation sequence of [6, 7, 19, 20], under boundedness of Y (or moments of order > 4) and show strong consistency of the estimation sequence consisting of the arithmetic means in the general case $\boldsymbol{E}\{Y^2\} < \infty$ (Section 4.2).

In Section 4.3 for the estimation of the local variance function σ^2 on the basis of data (4.1), we propose an estimation sequence (σ_n^2) of local averaging, namely partitioning, type. It is a modification of the (global) residual variance estimator and uses again first and second nearest neighbors instead of an estimation of m. We show strong L_2-consistency, i.e., $\int |\sigma_n^2(x) - \sigma^2(x)|^2 \mu(dx) \to 0$ *a.s.*, under mere boundedness of Y (μ denoting the distribution of X). Further, in the case of finite p-th moment of Y for some $p > 4$, $\boldsymbol{E}\{\int |\sigma_n^2(x) - \sigma^2(x)|^2 \mu(dx)\} \to 0$ is established.

In Section 4.4 we investigate the convergence rate of the local variance estimator with cubic partitioning. We assume that m and σ^2 belong to Lipschitz classes of type α and β, respectively, with $0 < \alpha \leq 1$, $0 < \beta \leq 1$, i.e., they are Hölder continuous with exponents α and β, respectively ($\alpha = 1$ or $\beta = 1$ means Lipschitz continuity). Then, with cube side length $h_n \sim n^{-1/(2\beta+d)}$,

$$\boldsymbol{E}\left\{\int |\sigma_n^2(x) - \sigma^2(x)|^2 \mu(dx)\right\} = O\left(\max\left\{n^{-4\alpha/d}, n^{-2\beta/(2\beta+d)}\right\}\right).$$

Only uniform boundedness of the conditional fourth moment of Y and boundedness of X, especially no density condition on X, are assumed. Thus in the case of random design, for Hölder continuity the partitioning variance estimate of simple structure with first and second nearest neighbors (without plug-in) yields the same convergence order as in [2] for regular fixed design. If $4\alpha/d \geq 2\beta/(2\beta+d)$, i.e., $\alpha \geq \beta d/(2(2\beta+d))$, one has the same convergence rate as in the case of known m, i.e., in the case of classic partitioning regression estimation with dependent variable $(Y - m(X))^2$. For the class of Hölder continuous functions σ^2 with exponent $\beta \leq 1$ this convergence rate $n^{-2\beta/(2\beta+d)}$ is optimal see p. 37, Theorem 3.2, p. 66, Theorem 4.3 with proof [11], i.e., the sequence $n^{-2\beta/(2\beta+d)}$ is the lower minimax rate and is achieved, namely by partitioning estimates. It seems possible that by a modification of partitioning the boundedness assumption on X can be relaxed to a moment condition (compare [17]).

4.2 Residual Variance Estimation

In the literature different paradigms how to construct nonparametric estimates are treated. Beside the least squares approach, local averaging paradigms are used, especially kernel estimates, partitioning estimates and k-th nearest neighbor estimates. A reference is [11].
For given $i \in \{1,\dots,n\}$, the first nearest neighbor of X_i among $X_1,\dots,X_{i-1}$, $X_{i+1},\dots,X_n$ is denoted by $X_{[N,1]}$ with

$$N[i,1] := N_n[i,1] := \underset{1\leq j\leq n,\ j\neq i}{\arg\min}\ \rho(X_i, X_j), \tag{4.4}$$

here ρ is a metric (typically the Euclidean one) in $\mathbb{R}^d$. The k-th nearest neighbor of X_i among $X_1,\dots,X_{i-1},X_{i+1},\dots,X_n$ is defined as $X_{N[i,k]}$ via generalization of definition (4.4):

$$N[i,k] := N_n[i,k] := \underset{1\leq j\leq n,\ j\neq i,\ j\notin\{N[i,1],\dots,N[i,k-1]\}}{\arg\min}\ \rho(X_i, X_j), \tag{4.5}$$

by removing the preceding neighbors. If ties occur, a possibility to break them is given by taking the minimal index or by adding independent components Z_i, uniformly distributed on $[0,1]$, to the observation vectors X_i see pp. 86, 87 [11]. The latter possibility of tie-breaking allow us to assume throughout the paper that ties occur with probability zero.
Hence, we get a reorder of the data according to increasing values of the distance of the variable X_j $(j \in \{1,\dots,n\} \setminus \{i\})$ from the variable X_i $(i = 1,\dots,n)$. Correspondingly to that, we get also a new order for the variables Y_j :

$$\left(X_{N[i,1]},\ Y_{N[i,1]}\right),\ \dots,\left(X_{N[i,k]},\ Y_{N[i,k]}\right),\ \dots,\left(X_{N[i,n-1]},\ Y_{N[i,n-1]}\right).$$

In the following $N[i,1]$ and $N[i,2]$ will be used.
For the residual variance V, [6, 7] introduced and [19, 20] analyzed (and generalized) the estimator

$$V_n = \frac{1}{n}\sum_{i=1}^{n}\left(Y_i - Y_{N[i,1]}\right)\left(Y_i - Y_{N[i,2]}\right), \tag{4.6}$$

in view of square mean consistency and rate of convergence.
We shall establish strong consistency.

Theorem 4.1. *If $|Y| \le L$ for some $L \in R_+$, then*

$$V_n \to V \quad a.s. \quad (n \to \infty).$$

(Asymptotic unbiasedness of the estimator of the residual variance).

The proof is based on the McDiarmid inequality see, e.g., Theorem A.2 [11] and properties of nearest neighbors (see [11], Lemma 6.1 and Corollary 6.1 together with Lemma 6.3). We state them in the following lemmas.

Lemma 4.1. *(McDiarmid inequality) Let $Z_1, \ldots, Z_n$ be independent random variables taking values in a set A and assume that $f : A^n \to \mathbb{R}$ satisfies*

$$\sup_{z_1,\ldots,z_n,z_i' \in A} |f(z_1,\ldots,z_n) - f(z_1,\ldots,z_{i-1},z_i',z_{i+1},\ldots,z_n)| \le c_i,$$

$1 \le i \le n$. Then, for all $\epsilon > 0$,

$$\boldsymbol{P}\{f(Z_1,\ldots,Z_n) - \boldsymbol{E}f(Z_1,\ldots,Z_n) \ge \epsilon\} \le e^{-\frac{2\epsilon^2}{\sum_{i=1}^n c_i^2}},$$

and

$$\boldsymbol{P}\{\boldsymbol{E}f(Z_1,\ldots,Z_n) - f(Z_1,\ldots,Z_n) \ge \epsilon\} \le e^{-\frac{2\epsilon^2}{\sum_{i=1}^n c_i^2}}.$$

Lemma 4.2. *If $k_n/n \to 0$, then*

$$\|X_{N[1,k_n]} - X_1\| \to 0 \quad a.s.$$

Lemma 4.3. *Under the assumption that ties occur with probability zero,*

a)

$$\sum_{i=1}^{n} 1_{\{X \text{ is among the } k \text{ nearest neighbors of } X_i \text{ in } \{X_1,\ldots,X_{i-1},X,X_{i+1},\ldots,X_n\}\}} \le k\gamma_d \quad a.s. \quad (k \le n),$$

b) for any integrable function f and any $k \leq n-1$,

$$\sum_{j=1}^{k} \boldsymbol{E}\{|f(X_{N[1,j]})|\} \leq k\gamma_d \boldsymbol{E}\{|f(X_1)|\},$$

Here $\gamma_d < \infty$ depends only on d.

Proof of Theorem 4.1. In the first step we show

$$\boldsymbol{E}V_n \to V \tag{4.7}$$

(asymptotic unbiasedness), using only square integrability of Y, compare proof of Theorem 2.2 [20].
With the notations

$$\begin{aligned} b_{i,j} &= m(X_i) - m(X_j) \\ r_i &= Y_i - m(X_i), \end{aligned} \tag{4.8}$$

we can write, according to [19, 20]:

$$\begin{aligned} &\boldsymbol{E}\left\{\left(Y_i - Y_{N[i,1]}\right)\left(Y_i - Y_{N[i,2]}\right)\right\} = \\ &\boldsymbol{E}\left\{b_{i,N[i,1]}\left(r_i - r_{N[i,2]}\right)\right\} + \boldsymbol{E}\left\{b_{i,N[i,2]}\left(r_i - r_{N[i,1]}\right)\right\} \\ &+\boldsymbol{E}\left\{\left(r_i - r_{N[i,1]}\right)\left(r_i - r_{N[i,2]}\right)\right\} + \boldsymbol{E}\left\{b_{i,N[i,1]}b_{i,N[i,2]}\right\}. \end{aligned}$$

As shown in [19] and [20] via conditioning with respect to $X_1, \ldots, X_n$,

$$\boldsymbol{E}\left\{b_{i,N[i,1]}\left(r_i - r_{N[i,2]}\right)\right\} = \boldsymbol{E}\left\{b_{i,N[i,2]}\left(r_i - r_{N[i,1]}\right)\right\} = 0,$$

and

$$\boldsymbol{E}\left\{\left(r_i - r_{N[i,1]}\right)\left(r_i - r_{N[i,2]}\right)\right\} = \boldsymbol{E}\{r_i^2\} = \boldsymbol{E}\{(Y_i - m(X_i))^2\} = V.$$

Further

$$\left|\boldsymbol{E}\left\{b_{i,N[i,1]}b_{i,N[i,2]}\right\}\right| \leq \boldsymbol{E}\left\{|m(X_i) - m(X_{N[i,1]})||m(X_i) - m(X_{N[i,2]})|\right\}.$$

Thus, because the X_i's are identically distributed,

$$\begin{aligned} |\boldsymbol{E}V_n - V| &\leq \boldsymbol{E}\left\{|m(X_1) - m(X_{N[1,1]})||m(X_1) - m(X_{N[1,2]})|\right\} \\ &\leq \frac{1}{2}\boldsymbol{E}\left\{|m(X_1) - m(X_{N[1,1]})|^2\right\} + \frac{1}{2}\boldsymbol{E}\left\{|m(X_1) - m(X_{N[1,2]})|^2\right\}. \end{aligned}$$

Because the set of continuous functions on $\mathbb{R}^d$ with compact support is dense in $L_2(\mu)$ (see, e.g., [5], Chapter 4, Section 8.19, or [11], Theorem A.1), for an arbitrary $\epsilon > 0$ one can choose a continuous function $\widetilde{m}$ with compact support such that $\boldsymbol{E}\{|m(X_1) - \widetilde{m}(X_1)|^2\} \le \epsilon$. Then

$$\begin{aligned} &\boldsymbol{E}\{|m(X_1) - m(X_{N[1,1]})|\} \\ &\le 3\boldsymbol{E}\{|(m-\widetilde{m})(X_1)|^2\} + 3\boldsymbol{E}\{|(m-\widetilde{m})(X_{N[1,1]})|^2\} \\ &\quad +3\boldsymbol{E}\{|(\widetilde{m}(X_1) - \widetilde{m}(X_{N[1,1]})|^2\}. \end{aligned}$$

By Lemma 4.2 (with $k_n = 1$) and continuity of $\widetilde{m}$, one has

$$\widetilde{m}(X_{N[1,1]}) \to \widetilde{m}(X_1) \quad a.s.,$$

thus, by boundedness of $\widetilde{m}$,

$$\boldsymbol{E}\{|\widetilde{m}(X_1) - \widetilde{m}(X_{N[1,1]})|^2\} \to 0.$$

Further, by Lemma 4.3b,

$$\begin{aligned} &\boldsymbol{E}\{|(m-\widetilde{m})(X_{N[1,1]})|^2\} \\ &\le \gamma_d \boldsymbol{E}\{|(m-\widetilde{m})(X_1)|\} \le \gamma_d \epsilon. \end{aligned}$$

Therefore

$$\limsup_{n\to\infty} \boldsymbol{E}\{|m(X_1) - m(X_{N[1,1]})|^2\} \le 3(1+\gamma_d)\epsilon,$$

thus

$$\boldsymbol{E}\{|m(X_1) - m(X_{N[1,1]})|^2\} \to 0.$$

Analogously one obtains $\boldsymbol{E}\{|m(X_1) - m(X_{N[1,2]})|^2\} \to 0$. Thus

$$\boldsymbol{E}\{|m(X_1) - m(X_{N[1,1]})||m(X_1) - m(X_{N[1,2]})|\} \to 0, \tag{4.9}$$

and (4.7) is obtained.
In the second step we show

$$V_n - \boldsymbol{E}V_n \to 0 \quad a.s. \tag{4.10}$$

Set

$$T_n := \sum_{i=1}^{n} (Y_i - Y_{N[i,1]})(Y_i - Y_{N[i,2]}).$$

Now in view of an application of Lemma 4.1, let $(X_1, Y_1), \ldots,$ (X_n, Y_n), $(X_1', Y_1'), \ldots, (X_n', Y_n')$ be independent and identically distributed $(d+1)$-dimensional random vectors. For fixed $j \in \{1, \ldots, n\}$ replace (X_j, Y_j)

by (X_j', Y_j'), which leads to $T_{n,j}$. Noticing $|Y_i| \le L$, we have

$$|T_n - T_{n,j}| \le 8L^2 + 8L^2 \cdot 2 \cdot 2\gamma_d = 8(1 + 4\gamma_d)L^2, \tag{4.11}$$

where the first term of the right-hand side results from addend $i = j$ and the second term results from addends $i \in \{1, \ldots, n\} \setminus \{j\}$, because replacement of X_j by X_j' has an influence on the first and second nearest neighbors of some, but at most $2\gamma_d$ (by Lemma 4.3 a), of the random vectors $X_1, \ldots, X_{j-1}$, $X_{j+1}, \ldots, X_n$. By Lemma 4.1, for each $\epsilon > 0$ we obtain

$$\begin{aligned} & \boldsymbol{P}\{|V_n - \boldsymbol{E}V_n| \ge \epsilon\} \\ &= \boldsymbol{P}\{|T_n - \boldsymbol{E}T_n| \ge \epsilon n\} \\ &\le 2e^{-2\epsilon^2 n^2 / n(8(1+4\gamma_d)L^2)^2}, \end{aligned} \tag{4.12}$$

thus (4.10) by the Borel-Cantelli lemma.
(4.7) and (4.10) yield the assertion. □

Remark 4.1. A truncation argument similar to that in the proof of Theorem 4.2 below yields that boundedness of Y in Theorem 4.1 can be relaxed to the moment condition $\boldsymbol{E}\{|Y|^p\} < \infty$ for some $p > 4$, because (4.12) then holds for L replaced by $n^{1/p}$.

The following theorem states that the boundedness assumption in Theorem 4.1 on Y can be omitted if for estimation of V the sequence $(W_n) = ((V_1 + \cdots + V_n)/n)$ of arithmetic means instead of (V_n) is used. It remains an open problem whether $V_n \to V$ *a.s.* if $\boldsymbol{E}\{Y^2\} < \infty$.

Theorem 4.2. *In the general case* $\boldsymbol{E}\{Y^2\} < \infty$,

$$W_n := \frac{V_1 + \cdots + V_n}{n} \to V \quad a.s.$$

Proof. For a real random variable U we set

$$U^{[c]} := U 1_{\{|U| \le c\}} + c 1_{\{U > c\}} - c 1_{\{U < -c\}}, \quad c > 0.$$

First we show

$$\frac{1}{n} \sum_{i=1}^{n} \left(Y_i - Y_{N[i,1]}\right)\left(Y_i - Y_{N[i,2]}\right) - \frac{1}{n} \sum_{i=1}^{n} V_{n,i} \to 0 \quad a.s.,$$

where

$$V_{n,i} := \left(Y_i^{[\sqrt{n}]} - Y_{N[i,1]}^{[\sqrt{n}]}\right)\left(Y_i^{[\sqrt{n}]} - Y_{N[i,2]}^{[\sqrt{n}]}\right).$$

Because $\boldsymbol{E}\{Y^2\} < \infty$, *a.s.* $Y_i = Y_i^{[\sqrt{i}]}$ for i sufficiently large, say, $i \geq M$ (random). For $i \in \{M, M+1, \ldots, n\}$, *a.s.* $Y_i = Y_i^{[\sqrt{n}]}$. By Lemma 4.3a, for $p \in \{1, \ldots, M\}$ one has $N[i,1] = p$ for at most γ_d indices $i \in \{1, \ldots, n\}$ and $N[i,2] = p$ for at most $2\gamma_d$ indices $i \in \{1, \ldots, n\}$. Thus *a.s.*

$$\left(Y_i - Y_{N[i,1]}\right)\left(Y_i - Y_{N[i,2]}\right) \neq \left(Y_i^{[\sqrt{n}]} - Y_{N[i,1]}^{[\sqrt{n}]}\right)\left(Y_i^{[\sqrt{n}]} - Y_{N[i,2]}^{[\sqrt{n}]}\right)$$

for at most $(1+3\gamma_d)M$ indices $i \in \{1, \ldots, n\}$, which yields the assertion. Therefore it suffices to show

$$\frac{1}{n}\sum_{l=1}^{n}\left(\frac{1}{l}\sum_{i=1}^{l} V_{l,i}\right) \to V \quad a.s. \tag{4.13}$$

In the second step we show

$$\frac{1}{n}\sum_{i=1}^{n} \boldsymbol{E}V_{n,i} \to V. \tag{4.14}$$

With $m^{(n)}(x) := \boldsymbol{E}\{Y^{[\sqrt{n}]} | X = x\}$ we have

$$\begin{aligned}
&\frac{1}{n}\sum_{i=1}^{n} \boldsymbol{E}V_{n,i} = \boldsymbol{E}V_{n,1} \\
&= \boldsymbol{E}\left\{(Y^{[\sqrt{n}]} - m^{(n)}(X))^2\right\} \\
&+ \boldsymbol{E}\left\{\left(m^{(n)}(X_1) - m^{(n)}(X_{N[1,1]})\right)\left(m^{(n)}(X_1) - m^{(n)}(X_{N[1,2]})\right)\right\},
\end{aligned}$$

the latter according to [19, 20]. By $\boldsymbol{E}\{Y^2\} < \infty$ and the dominated convergence theorem, $\int |m^{(n)}(x) - m(x)|^2 \mu(dx) \to 0$ and thus $\boldsymbol{E}\{(Y^{[\sqrt{n}]} - m^{(n)}(X))^2\} \to V$. Further m and also $m^{(n)}$ can be approximated by a continuous function $\tilde{m}$ with compact support such that for each $\epsilon > 0$ an index $n_0(\epsilon)$ exists with $\boldsymbol{E}\{|m(X) - \tilde{m}(X)|^2\} \leq \epsilon$ and also

$$\boldsymbol{E}\{|m^{(n)}(X) - \tilde{m}(X)|^2\} \leq \epsilon \text{ for } n \geq n_0(\epsilon).$$

Then we obtain

$$\begin{aligned}
&\boldsymbol{E}\left\{\left|m^{(n)}(X_1) - m^{(n)}(X_{N[1,1]})\right|^2\right\} \\
&\leq 3\boldsymbol{E}\{|(m^{(n)} - \tilde{m})(X_1)|^2\} +
\end{aligned}$$

$$3\boldsymbol{E}\{|(m^{(n)} - \widetilde{m})(X_{N[1,1]})|^2\} + 3\boldsymbol{E}\{|\widetilde{m}(X_1) - \widetilde{m}(X_{N[1,1]})|^2\}$$
$$\leq 3\epsilon + 3\gamma_d\epsilon + o(1),$$

the latter as in the proof of Theorem 4.1. Therefore

$$\boldsymbol{E}\left\{\left|m^{(n)}(X_1) - m^{(n)}(X_{N[1,1]})\right|^2\right\} \to 0$$

and correspondingly

$$\boldsymbol{E}\left\{\left|m^{(n)}(X_1) - m^{(n)}(X_{N[1,2]})\right|^2\right\} \to 0,$$

thus

$$\boldsymbol{E}\left\{\left(m^{(n)}(X_1) - m^{(n)}(X_{N[1,1]})\right)\left(m^{(n)}(X_1) - m^{(n)}(X_{N[1,2]})\right)\right\}$$

$\to 0$, and (4.14) is obtained as well as

$$\frac{1}{n}\sum_{l=1}^{n}\left(\frac{1}{l}\sum_{i=1}^{l}\boldsymbol{E}V_{l,i}\right) \to V. \tag{4.15}$$

In the second step we show

$$\frac{1}{n}\sum_{l=1}^{n}\left(\frac{1}{l}\sum_{i=1}^{l}(V_{l,i} - \boldsymbol{E}V_{l,i})\right) \to 0 \quad a.s. \tag{4.16}$$

It suffices to show

$$\sum \frac{\boldsymbol{Var}\left\{\sum_{i=1}^{n} V_{n,i}\right\}}{n^3} < \infty, \tag{4.17}$$

for this implies

$$\sum \frac{1}{n}\left(\frac{1}{n}\sum_{i=1}^{n}(V_{n,i} - \boldsymbol{E}V_{n,i})\right)^2 < \infty \quad a.s.$$

and, by the Cauchy-Schwarz inequality and the Kronecker lemma,

$$\left|\frac{1}{n}\sum_{l=1}^{n}\left(\frac{1}{l}\sum_{i=1}^{l}(V_{l,i} - \boldsymbol{E}V_{l,i})\right)\right|^2$$

$$\leq \frac{1}{n}\sum_{l=1}^{n}\left|\frac{1}{l}\sum_{i=1}^{l}(V_{l,i}-\boldsymbol{E}V_{l,i})\right|^2 \to 0 \quad a.s.$$

We shall show

$$\boldsymbol{Var}\left\{\sum_{i=1}^{n}V_{n,i}\right\} \leq cn\boldsymbol{E}\left\{\left(Y^{[\sqrt{n}]}\right)^4\right\},\ n \in \mathbb{N} \tag{4.18}$$

for a suitable finite constant c. This, together with $\boldsymbol{E}\{Y^2\} < \infty$, implies (4.17), because, as is well known (see, e.g., Section 17.3 [21]), $\boldsymbol{E}|U| < \infty$ for a real variable U implies $\sum \boldsymbol{E}\left\{\left(U^{[n]}\right)^2\right\}/n^2$
$< \infty$.
We prove (4.18) by using the Efron-Stein inequality in [32] version (compare also Theorem A.3 [11]).
Let $n \geq 2$ be fixed. Replacement of (X_j, Y_j) by (X_j', Y_j') for fixed $j \in \{1,\dots,n\}$ (where $(X_1,Y_1),\dots(X_n,Y_n),(X_1',Y_1'),\dots,$
(X_n',Y_n') are independent and identically distributed) leads from $T_n := \sum_{i=1}^{n}V_{n,i}$, $N[j,1]$ and $N[j,2]$ to $T_{n,j}$, $N'[j,1]$ and $N'[j,2]$, respectively.
We obtain

$$|T_n - T_{n,j}| \leq A_{n,j} + B_{n,j} + C_{n,j} + D_{n,j} + E_{n,j} + F_{n,j}$$

where with $Z_i = Y_i^{[\sqrt{n}]}$, $Z_j' = Y_j'^{[\sqrt{n}]}$, $Z = Y^{[\sqrt{n}]}$

$$A_{n,j} = \sum_{\substack{l,\ q\in\{1,\dots,n\}\setminus\{j\}\\ l\neq q}} |Z_j - Z_l||Z_j - Z_q| 1_{\{N[j,1]=l\}} 1_{\{N[j,2]=q\}},$$

$$B_{n,j} = \sum_{\substack{l,\ q\in\{1,\dots,n\}\setminus\{j\}\\ l\neq q}} |Z_j' - Z_l||Z_j' - Z_q| 1_{\{N'[j,1]=l\}} 1_{\{N'[j,2]=q\}},$$

$$C_{n,j} = \sum_{\substack{i,\ q\in\{1,\dots,n\}\setminus\{j\}\\ i\neq q}} |Z_i - Z_j||Z_i - Z_q| 1_{\{N[i,1]=j\}} 1_{\{N[i,2]=q\}},$$

$$D_{n,j} = \sum_{\substack{i,\ q\in\{1,\dots,n\}\setminus\{j\}\\ i\neq q}} |Z_i - Z_j'||Z_i - Z_q| 1_{\{N'[i,1]=j\}} 1_{\{N'[i,2]=q\}},$$

$$E_{n,j} = \sum_{\substack{i,\ l\in\{1,\dots,n\}\setminus\{j\}\\ i\neq l}} |Z_i - Z_l||Z_i - Z_j| 1_{\{N[i,1]=l\}} 1_{\{N[i,2]=j\}},$$

$$F_{n,j} = \sum_{\substack{i,\ l \in \{1,\dots,n\} \setminus \{j\} \\ i \neq l}} |Z_i - Z_l||Z_i - Z'_j| 1_{\{N'[i,1]=l\}} 1_{\{N'[i,2]=j\}}.$$

Thus

$$|T_n - T_{n,j}|^2 \le 6(A_{n,j}^2 + B_{n,j}^2 + C_{n,j}^2 + D_{n,j}^2 + E_{n,j}^2 + F_{n,j}^2).$$

By the Cauchy-Schwarz inequality applied to the sums defining $A_{n,j}, \dots,$ $F_{n,j}$ and by the inequality $|a-b|^2|a-c|^2 \le 8(a^4 + b^4 + c^4)$ we obtain

$$\begin{aligned} &\boldsymbol{E} \sum_{j=1}^{n} |T_n - T_{n,j}|^2 \\ &\le const \boldsymbol{E} \sum_{\substack{j,\ l,\ q \in \{1,\dots,n\} \\ j \neq l,\ l \neq q,\ q \neq j}} (Z_j^4 + Z_l^4 + Z_q^4) 1_{\{N[j,1]=l\}} 1_{\{N[j,2]=q\}}. \end{aligned}$$

As to the term concerning Z_j^4 we sum with respect to l and q and for the corresponding expected final sum we obtain the bound $n\boldsymbol{E}\left\{Z^4\right\}$. As to the term Z_l^4 we sum with respect to q, then with respect to j using Lemma 4.3a, and for the corresponding expected final sum we obtain the bound $\gamma_d n\boldsymbol{E}\left\{Z^4\right\}$. As to the term Z_q^4 we sum with respect l, then with respect to j using Lemma 4.3b and for the corresponding expected final sum we obtain the bound $2\gamma_d n\boldsymbol{E}\left\{Z^4\right\}$.
Therefore, by the Efron-Stein inequality,

$$\boldsymbol{Var}(T_n) \le c^* n \boldsymbol{E}\left\{\left(Y^{[\sqrt{n}]}\right)^4\right\},$$

i.e., (4.18). Thus (4.16) is obtained, which together with (4.15) implies (4.13). □

4.3 Local Variance Estimation: Strong and Weak Consistency

V_n in (4.6) as an estimator of $V = \boldsymbol{E}\{(Y - m(X))^2\}$ was treated in Section 4.2. In this section our aim is to give an estimator of the local variance function σ^2 in (4.2). Recall the relation between the residual and the local variance function in (4.3).

Our proposal for an appropriate estimator of σ^2 is

$$\sigma_n^2(x) := \frac{\sum_{i=1}^n (Y_i - Y_{N[i,1]})(Y_i - Y_{N[i,2]}) 1_{A_n(x)}(X_i)}{\sum_{i=1}^n 1_{A_n(x)}(X_i)}, \; x \in \mathbb{R}^d \tag{4.19}$$

with $0/0 := 0$, where $\mathcal{P}_n = \{A_{n,1}, A_{n,2}, \dots\}$ is a partition of $\mathbb{R}^d$ consisting of Borel sets $A_{n,j} \subset \mathbb{R}^d$, and where the notation $A_n(x)$ is used for the $A_{n,j}$ containing x. In this sense we localize the global expression in V_n by local averaging, in particular by partitioning. Analogously a kernel type estimator could be treated. The next theorem deals with strong consistency of the local variance estimator under the assumption that Y is bounded. It comprehends the special case of cubic partitioning.

Theorem 4.3. *Let* $(\mathcal{P}_n)_{n\in\mathbb{N}}$ *with* $\mathcal{P}_n = \{A_{n,1}, A_{n,2}, \dots\}$ *be a sequence of partitions of* $\mathbb{R}^d$ *such that for each sphere* S *centered at the origin*

$$\lim_{n\to\infty} \max_{j: A_{n,j}\cap S \neq \emptyset} \mathit{diam}\, A_{n,j} \to 0 \tag{4.20}$$

and, for some $\rho = \rho(S) \in (0, \frac{1}{2})$,

$$\#\{j : A_{n,j} \cap S \neq \emptyset\} \sim n^\rho. \tag{4.21}$$

Assume $|Y| \leq L$ *for some* $L \in \mathbb{R}_+$. *Then*

$$\int |\sigma_n^2(x) - \sigma^2(x)|^2 \mu(dx) \to 0 \quad a.s.$$

(Strong consistency of the partitioning estimator of the local variance)

Proof. Because of the boundedness assumption it suffices to prove

$$\int |\sigma_n^2(x) - \sigma^2(x)| \mu(dx) \to 0 \quad a.s. \tag{4.22}$$

Set

$$\sigma_n^{2*}(x) := \frac{\sum_{i=1}^n (Y_i - Y_{N[i,1]})(Y_i - Y_{N[i,2]}) 1_{A_n(x)}(X_i)}{n\mu(A_n(x))}. \tag{4.23}$$

In the first step we show that for each sphere S centered at 0

$$\boldsymbol{E}\left\{\int_S |\sigma^2(x) - \sigma_n^{2*}(x)| \mu(dx)\right\} \to 0. \tag{4.24}$$

We have

$$\begin{aligned}
&\mathbf{E}\left\{\int_S |\sigma^2(x) - \sigma_n^{2*}(x)|\mu(dx)\right\} \\
&\le \int_S |\sigma^2(x) - \mathbf{E}\sigma_n^{2*}(x)|\mu(dx) \\
&\quad + \mathbf{E}\left\{\int_S |\mathbf{E}\sigma_n^{2*}(x) - \sigma_n^{2*}(x)|\mu(dx)\right\} \\
&\le K_n + M_n.
\end{aligned}$$

We show $K_n \to 0$. According to [19, 20] we have

$$\begin{aligned}
&\mathbf{E}\{(Y_1 - Y_{N[1,1]})(Y_1 - Y_{N[1,2]})|X_1 = z\} \\
&= \sigma^2(z) \\
&\quad + \mathbf{E}\{(m(X_1) - m(X_{N[1,1]}))(m(X_1) - m(X_{N[1,2]}))|X_1 = z\},
\end{aligned}$$

thus

$$\begin{aligned}
&\mathbf{E}\sigma_n^{2*}(x) \\
&= \int \frac{\sigma^2(z) 1_{A_n(x)}(z)}{\mu(A_n(x))}\mu(dz) \\
&\quad + \int \frac{\mathbf{E}\{(m(X_1) - m(X_{N[1,1]}))(m(X_1) - m(X_{N[1,2]}))|X_1 = z\} 1_{A_n(x)}(z)}{\mu(A_n(x))}\mu(dz).
\end{aligned}$$

Notice

$$\begin{aligned}
&\int \left[\int \frac{\mathbf{E}\{|m(X_1) - m(X_{N[1,1]})||m(X_1) - m(X_{N[1,2]})||X_1 = z\} 1_{A_n(x)}(z)}{\mu(A_n(x))}\mu(dz)\right]\mu(dx) \\
&= \int \left[\int \frac{\mathbf{E}\{|m(X_1) - m(X_{N[1,1]})||m(X_1) - m(X_{N[1,2]})||X_1 = z\} 1_{A_n(z)}(x)}{\mu(A_n(x))}\mu(dx)\right]\mu(dz) \\
&\le \mathbf{E}\{|m(X_1) - m(X_{N[1,1]}||m(X_1) - m(X_{N[1,2]})|\} \\
&\to 0
\end{aligned}$$

by (4.9). Moreover,

$$\int \left|\sigma^2(x) - \int \frac{\sigma^2(z) 1_{A_n(x)}(z)}{\mu(A_n(x))}\mu(dz)\right| \mu(dx) \to 0.$$

For, because of $\int \sigma^2(x)\mu(dx) < \infty$, as in the proof of Theorem 4.1 for each $\epsilon > 0$ we can choose a continuous function $\widetilde{\sigma}^2$ with compact support such that

$$\int |\sigma^2(x) - \widetilde{\sigma}^2(x)|\mu(dx) < \epsilon,$$

further

$$\int \left| \int \frac{\sigma^2(z) 1_{A_n(x)}(z)}{\mu(A_n(x))} \mu(dz) - \int \frac{\widetilde{\sigma}^2(z) 1_{A_n(x)}(z)}{\mu(A_n(x))} \mu(dz) \right| \mu(dx)$$
$$\leq \int \left| \sigma^2(z) - \widetilde{\sigma}^2(z) \right| \mu(dz) < \epsilon,$$

and we notice

$$\int_S \left| \widetilde{\sigma}^2(x) - \int \frac{\widetilde{\sigma}^2(z) 1_{A_n(x)}(z)}{\mu(A_n(x))} \mu(dz) \right| \mu(dx) \to 0$$

because of uniform continuity of $\widetilde{\sigma}$ and (4.20). Therefore $K_n \to 0$.
Now M_n will be treated. Set $J_n := \{j : A_{n,j} \cap S \neq \emptyset\}$ and $l_n := \# J_n$.

$$\begin{aligned}
M_n = & \sum_{j \in J_n} \boldsymbol{E} \left\{ \int_{A_{n,j}} \left| \frac{\sum_{i=1}^n (Y_i - Y_{N[i,1]})(Y_i - Y_{N[i,2]}) 1_{A_{n,j}}(X_i)}{n\mu(A_{n,j})} \right. \right. \\
& \left. \left. - \boldsymbol{E} \frac{\sum_{i=1}^n (Y_i - Y_{N[i,1]})(Y_i - Y_{N[i,2]}) 1_{A_{n,j}}(X_i)}{n\mu(A_{n,j})} \right| \mu(dx) \right\} \\
\leq & \frac{1}{n} \sum_{j \in J_n} \boldsymbol{E} \left\{ \left| \sum_{i=1}^n (Y_i - Y_{N[i,1]})(Y_i - Y_{N[i,2]}) 1_{A_{n,j}}(X_i) \right. \right. \\
& \left. \left. - \boldsymbol{E} \sum_{i=1}^n (Y_i - Y_{N[i,1]})(Y_i - Y_{N[i,2]}) 1_{A_{n,j}}(X_i) \right| \right\} \\
\leq & \frac{1}{n} \sum_{j \in J_n} \sqrt{\boldsymbol{Var} \left\{ \sum_{i=1}^n (Y_i - Y_{N[i,1]})(Y_i - Y_{N[i,2]}) 1_{A_{n,j}}(X_i) \right\}} \\
\leq & \frac{l_n}{n} \sqrt{\frac{n}{2} (8L^2 + 8L^2 \cdot 2 \cdot 2\gamma_d)^2} \\
& \text{(by the Efron-Stein lemma and the derivation of (4.11))} \\
\leq & 4\sqrt{2}(1 + 4\gamma_d) L^2 \frac{l_n}{\sqrt{n}} \to 0 \quad \text{(by (4.21))}.
\end{aligned}$$

Thus (4.24) is obtained.
In the second step we show: for an arbitrary sphere S centered at 0, a constant $c > 0$ exists such that for each $\epsilon > 0$

$$\boldsymbol{P} \left\{ \int_S |\sigma^2(x) - \sigma_n^{2*}(x)| \mu(dx) > 2\epsilon \right\} \leq e^{-\epsilon^2 c n^{1-2\rho}} \tag{4.25}$$

for n sufficiently large, thus, by the Borel-Cantelli lemma,

$$\int_S |\sigma_n^{2*}(x) - \sigma^2(x)|\mu(dx) \to 0 \quad a.s. \tag{4.26}$$

We follow the argument in the proof of Lemma 23.2 in [11].
It holds

$$\begin{aligned} & |\sigma^2(x) - \sigma_n^{2*}(x)| \\ &= \boldsymbol{E}\left\{|\sigma^2(x) - \sigma_n^{2*}(x)|\right\} + (|\sigma^2(x) - \sigma_n^{2*}(x)| \\ & -\boldsymbol{E}\left\{|\sigma^2(x) - \sigma_n^{2*}(x)|\right\}). \end{aligned}$$

At the first term on the right-hand side, we notice (4.24). As to the second term, in view of an application of McDiarmid's inequality (Lemma 4.1) replacing (X_i, Y_i) by (X_i', Y_i') as in the proof of Theorem 4.1 leads from $\sigma_n^{2*}(x)$ to $\sigma_{n,j}^{2*}(x)$, $j \in \{1, \ldots, n\}$, where, correspondingly to (4.11),

$$|\sigma_n^{2*}(x) - \sigma_{n,j}^{2*}(x)| \le \frac{8(1 + 4\gamma_d)L^2}{n\mu(A_n(x))}.$$

Thus

$$\begin{aligned} & \left| \int_S |\sigma^2(x) - \sigma_n^{2*}(x)|\mu(dx) - \int_S |\sigma^2(x) - \sigma_{n,j}^{2*}(x)|\mu(dx) \right| \\ &= \left| \int_S \left(|\sigma^2(x) - \sigma_n^{2*}(x)| - |\sigma^2(x) - \sigma_{n,j}^{2*}(x)| \right) \mu(dx) \right| \\ &\le \int_S |\sigma_n^2(x) - \sigma_{n,j}^{2*}(x)|\mu(dx) \quad (j = 1, \ldots, n) \\ & \text{(due to the triangle inequality } |a - b| \ge ||a| - |b||) \\ &\le \frac{8(1 + 4\gamma_d)L^2}{n} \int_S \frac{1}{\mu(A_n(x))}\mu(dx) \\ &\le \frac{8(1 + 4\gamma_d)L^2}{n} l_n, \end{aligned}$$

where $l_n := \#\{j : A_{n,j} \cap S \neq \emptyset\}$. Now, using (4.21) once more as well as Lemma 4.1, for arbitrary $\epsilon > 0$

$$\begin{aligned} & \boldsymbol{P}\left\{ \int_S |\sigma^2(x) - \sigma_n^{2*}(x)|\mu(dx) - \boldsymbol{E} \int_S |\sigma^2(x) - \sigma_n^{2*}(x)|\mu(dx) > \epsilon \right\} \\ &\le e^{-2\epsilon^2 / \left(n \frac{[8(1+4\gamma_d)L^2]^2}{n^2} l_n^2 \right)} \le^{-\epsilon^2 c n^{1-2\rho}}. \end{aligned}$$

with some $c > 0$. Therefore, because of $\int_S \boldsymbol{E}|\sigma^2(x) - \sigma_n^{2*}(x)|\mu(dx) < \epsilon$ for n large enough, (4.25) is obtained.
In the third step we show (4.22). Because Y is bounded, for an arbitrary $\epsilon > 0$ we can choose a sphere S centered at 0, such that

$$\int_{S^c} |\sigma_n^2(x) - \sigma^2(x)|\mu(dx) \le \epsilon.$$

Therefore it suffices to show $\int_S |\sigma_n^2(x) - \sigma^2(x)|\mu(dx) \to 0 \; a.s.$ for each sphere S centered at 0. We have

$$\begin{aligned}
&\int_S |\sigma_n^2(x) - \sigma^2(x)|\mu(dx) \\
&\le \int_S |\sigma_n^2(x) - \sigma_n^{2*}(x)|\mu(dx) + \int_S |\sigma_n^{2*}(x) - \sigma^2(x)|\mu(dx) \\
&\le G_n + D_n,
\end{aligned}$$

where $D_n \to 0 \; a.s.$ by (4.26). Now, concerning G_n, similarly to the argument in [11], p. 465, by (4.20) and (4.21),

$$\begin{aligned}
&\int_S |\sigma_n^{2*}(x) - \sigma_n^2(x)|\mu(dx) \\
&\le \int \left| \frac{\sum_{i=1}^n (Y_i - Y_{N[i,1]})(Y_i - Y_{N[i,2]}) 1_{A_n(x)}(X_i)}{n\mu(A_n(x))} \right. \\
&\quad \left. - \frac{\sum_{i=1}^n (Y_i - Y_{N[i,1]})(Y_i - Y_{N[i,2]}) 1_{A_n(x)}(X_i)}{\sum_{i=1}^n 1_{A_n(x)}(X_i)} \right| \mu(dx) \\
&\le 4L^2 \int \sum_{i=1}^n 1_{A_n(x)}(X_i) \left| \frac{1}{n\mu(A_n(x))} - \frac{1}{\sum_{i=1}^n 1_{A_n(x)}(X_i)} \right| \mu(dx) \\
&\le 4L^2 \int \left| \sum_{i=1}^n \frac{1_{A_n(x)}(X_i)}{n\mu(A_n(x))} - 1 \right| \mu(dx) \to 0 \quad a.s.,
\end{aligned}$$

and (4.22) is obtained. □

In the following we establish weak consistency of (σ_n^2) under a moment condition on Y. As is well-known the partitioning regression estimation sequence (m_n), defined analogously to (4.19) with

$$U_{n,i} := (Y_i - Y_{N[i,1]})(Y_i - Y_{N[i,2]}) \tag{4.27}$$

replaced by Y_i, is weakly universally consistent see Theorem 4.2 [11], i.e.,

$$\boldsymbol{E}\left\{\int |m_n(x) - m(x)|^2 \mu(dx)\right\} \to 0 \tag{4.28}$$

for each distribution of (X, Y) with $\boldsymbol{E}\{Y^2\} < \infty$. The corresponding property holds for kernel regression estimation, but not for piecewise linear partitioning and local linear regression estimation see Theorem 5.1 and p. 81, respectively [11]. The corresponding weak universal consistency result for (σ_n^2) would mean

$$\boldsymbol{E}\left\{\int |\sigma_n^2(x) - \sigma^2(x)|^2 \mu(dx)\right\} \to 0 \tag{4.29}$$

for each distribution of (X, Y) with $\boldsymbol{E}\{Y^4\} < \infty$. The following Theorem 3.2 establishes (4.29) under the slightly stronger moment condition

$$\boldsymbol{E}\left\{|Y|^{4p}\right\} < \infty \tag{4.30}$$

for some $p > 1$. The property $\boldsymbol{E}\{Y_i^k | X_1, \ldots, X_n\} = \boldsymbol{E}\{Y_i^k | X_i\}$ of Y_i $(i = 1, \ldots, n;\ k = 1, 2)$ used in the proof of (4.28) see p. 59 in context of p. 62 [11] obviously does not analogously hold for $U_{n,i}$. In the proof of Theorem 3.2 we avoid the obstacle by use of the Hölder's inequality for conditional expectations.

Theorem 4.4. *Let the sequence $(\mathcal{P}_n)_{n \in \mathbb{N}}$ of partitions of $\mathbb{R}^d$ satisfy (4.20) and (4.21). Assume existence of a $p > 1$ such that (4.30) holds. Then (4.29) holds.*

Proof. We first use an argument in the proof of the regression estimation Lemma 23.2 in Györfi et al. (2002). For $c > 0$ set

$$Y_i^{(c)} := \begin{cases} c & \text{if } Y > c \\ Y_i & \text{if } -c \le Y_i \le c \\ -c & \text{if } Y < -c. \end{cases}$$

In context of Y, define $Y^{(c)}$ correspondingly.
Set

$$U_{n,i}^{(c)} := (Y_i^{(c)} - Y_{N[i,1]}^{(c)})(Y_i^{(c)} - Y_{N[i,2]}^{(c)})$$

and define $\sigma_n^{2(c)}$ analogously to (4.19) replacing $U_{n,i}$ by $U_{n,i}^{(c)}$.
Set

$$\sigma^{2(c)}(x) := \boldsymbol{E}\{(Y^{(c)} - \boldsymbol{E}\{Y^{(c)} | X = x\})^2 | X = x\}, \quad x \in \mathbb{R}^d.$$

Then

$$\begin{aligned}
&\int |\sigma_n^2(x) - \sigma^2(x)|^2 \mu(dx) \\
&\leq 3 \left(\int |\sigma_n^2(x) - \sigma_n^{2(c)}(x)|^2 \mu(dx) + \int |\sigma_n^{2(c)}(x) - \sigma^{2(c)}(x)|^2 \mu(dx) \right. \\
&\quad \left. + \int |\sigma^{2(c)}(x) - \sigma^2(x)|^2 \mu(dx) \right).
\end{aligned}$$

By the dominated convergence theorem

$$\int |\sigma^{2(c)}(x) - \sigma^2(x)|^2 \mu(dx) \to 0 \quad (c \to \infty).$$

By Theorem 4.3 and the dominated convergence theorem

$$\boldsymbol{E}\left\{ \int |\sigma_n^2(x)^{(c)} - \sigma^{2(c)}(x)|^2 \mu(dx) \right\} \to 0 \quad (n \to \infty).$$

Therefore it suffices to show

$$\lim_{c\to\infty} \limsup_{n\to\infty} \boldsymbol{E}\left\{ \int |\sigma_n^2(x) - \sigma_n^{2(c)}(x)|^2 \mu(dx) \right\} = 0.$$

By Jensen's inequality

$$\begin{aligned}
&\int |\sigma_n^2(x) - \sigma_n^{2(c)}(x)|^2 \mu(dx) \\
&\leq \int \sum_{i=1}^{n} \frac{(U_{n,i} - U_{n,i}^{(c)})^2 1_{A_n(x)}(X_i)}{1 + \sum_{j\in\{1,\dots,n\}\setminus\{i\}} 1_{A_n(x)}(X_j)} \mu(dx).
\end{aligned}$$

Therefore and by symmetry it is enough to show

$$\lim_{c\to\infty} \limsup_{n\to\infty} n \int \boldsymbol{E}\left\{ \frac{(U_{n,1} - U_{n,1}^{(c)})^2 1_{A_n(x)}(X_1)}{1 + \sum_{j=2}^{n} 1_{A_n(x)}(X_j)} \mu(dx) \right\} = 0.$$

Without loss of generality assume that $q > 1$ which satisfies $\frac{1}{p} + \frac{1}{q} = 1$, is an integer.
We have, with suitable $c' \in \mathbb{R}_+$,

$$n \int \boldsymbol{E}\left\{ \frac{(U_{n,1} - U_{n,1}^{(c)})^2 1_{A_n(x)}(X_1)}{1 + \sum_{j=2}^{n} 1_{A_n(x)}(X_j)} \right\} \mu(dx)$$

$$= n \int \boldsymbol{E}\left\{ \boldsymbol{E}\left\{ \frac{(U_{n,1} - U_{n,1}^{(c)})^2 1_{A_n(x)}(X_1)}{1 + \sum_{j=2}^{n} 1_{A_n(x)}(X_j)} \middle| X_1 \right\} \right\} \mu(dx)$$

$$\leq n \int \boldsymbol{E}\left\{ 1_{A_n(x)}(X_1) \left(\boldsymbol{E}\left\{ |U_{n,1} - U_{n,1}^{(c)}|^{2p} | X_1 \right\} \right)^{1/p} \right\}$$

$$\left(\boldsymbol{E}\left\{ \frac{1}{(1 + \sum_{j=2}^{n} 1_{A_n(x)}(X_j))^q} \right\} \right)^{1/q} \mu(dx)$$

(by Hölder's inequality for conditional expectations and independence of the pair $(X_1, (X_2, \dots, X_n))$

$$\leq c'n \int \boldsymbol{E}\left\{ 1_{A_n(x)}(X_1) \boldsymbol{E}\left\{ |U_{n,i} - U_{n,i}^{(c)}|^{2p} | X_1 \right\}^{1/p} \right\} \frac{1}{n\mu(A_n(x))} \mu(dx)$$

(because the sum in the denominator is $b(n-1, \mu(A_n(x))$-distributed, compare Lemma 4.1 [11])

$$\leq c' \boldsymbol{E}\left\{ \left(\boldsymbol{E}\left\{ |U_{n,1} - U_{n,1}^{(c)}|^{2p} | X_1 \right\} \right)^{1/p} \right\}$$

(because of $\int (1_{A_n(x)}(t)/\mu(A_n(x))\mu(dx) = \int (1_{A_n(t)}(x)/\mu(A_n(t))\mu(dx) \leq 1$ for all $t \in \mathbb{R}^d$)

$$\leq c' \left(\boldsymbol{E}\left\{ \boldsymbol{E}\left\{ |U_{n,1} - U_{n,1}^{(c)}|^{2p} | X_1 \right\} \right\} \right)^{1/p}$$

(by Jensen's inequality)

$$= c' \left(\boldsymbol{E}\left\{ |U_{n,1} - U_{n,1}^{(c)}|^{2p} \right\} \right)^{1/p}.$$

Therefore it suffices to show

$$\lim_{c \to \infty} \limsup_{n \to \infty} \boldsymbol{E}\left\{ |U_{n,1} - U_{n,1}^{(c)}|^{2p} \right\} = 0. \tag{4.31}$$

We notice

$$\begin{aligned}
&|U_{n,1} - U_{n,1}^{(c)}| \\
&\leq |Y_1^2 - \left(Y_1^{(c)}\right)^2| + |Y_1||Y_{N[1,1]} - Y_{N[1,1]}^{(c)}| + |Y_1 - Y_1^{(c)}||Y_{N[1,1]}^{(c)}| \\
&\quad + |Y_{N[1,2]}||Y_1 - Y_1^{(c)}| + |Y_{N[1,2]} - Y_{N[1,2]}^{(c)}||Y_1^{(c)}| \\
&\quad + |Y_{N[1,1]}||Y_{N[1,2]} - Y_{N[1,2]}^{(c)}| + |Y_{N[1,1]} - Y_{N[1,1]}^{(c)}||Y_{N[1,2]}^{(c)}|,
\end{aligned}$$

and by use of the Cauchy-Schwarz inequality we obtain

$$\begin{aligned}
&\boldsymbol{E}\left\{|U_{n,1} - U_{n,1}^{(c)}|^{2p}\right\} \\
&\le c^* \boldsymbol{E}\{|Y_1^2 - \left(Y_1^{(c)}\right)^2|^{2p}\} \\
&+c^* \left[(\boldsymbol{E}\{|Y_1|^{4p}\})^{1/2} + (\boldsymbol{E}\{|Y_{N[1,1]}|^{4p}\})^{1/2} + (\boldsymbol{E}\{|Y_{N[1,2]}|^{4p}\})^{1/2}\right] \\
&\left[\left(\boldsymbol{E}\{|Y_{N[1,1]} - Y_{N[1,1]}^{(c)}|^{4p}\}\right)^{1/2} + \left(\boldsymbol{E}\{|Y_{N[1,2]} - Y_{N[1,2]}^{(c)}|^{4p}\}\right)^{1/2}\right. \\
&\left. + (\boldsymbol{E}\{|Y_1 - Y_2^c|^{4p}\})^{1/2}\right]
\end{aligned}$$

(with suitable $c^* \in \mathbb{R}_+$)

$$\begin{aligned}
&\le c^* \boldsymbol{E}\left\{|Y_1^2 - \left(Y_1^{(c)}\right)^2|^{2p}\right\} \\
&+c^{**} (\boldsymbol{E}\{|Y_1|^{4p}\})^{1/2} \left(\boldsymbol{E}\left\{|Y_1 - Y_1^{(c)}|^{4p}\right\}\right)^{1/2}
\end{aligned}$$

with suitable $c^{**} \in \mathbb{R}_+$ dependent on the dimension d, for all $n \in \mathbb{N}$, by Lemma 6.3 in Györfi et al. (2002). Letting $c \to \infty$ on the right-hand side yields (4.31) by the dominated convergence theorem. □

4.4 Rate of Convergence

In this section we establish a rate of convergence for the estimate of the local variance defined in (4.19).

Theorem 4.5. *Assume that X is bounded and*

$$\boldsymbol{E}\{Y^4|X = x\} \le \tau^4, \quad x \in \mathbb{R}^d, \tag{4.32}$$

($0 < \tau < \infty$). Moreover, assume the Hölder conditions

$$|\sigma^2(x) - \sigma^2(t)| \le D\|x - t\|^\beta, \quad x, t \in \mathbb{R}^d, \tag{4.33}$$

and

$$|m(x) - m(t)| \le D^*\|x - t\|^\alpha, \quad x, t \in \mathbb{R}^d, \tag{4.34}$$

with $0 < \alpha \le 1$, $0 < \beta \le 1$; D, $D^* \in \mathbb{R}_+$, *(*$\| \ \|$ *denoting the Euclidean norm).*
Let $\mathcal{P}_n$ *be a cubic partition of* $\mathbb{R}^d$ *with side length* h_n *of the cubes* $(n \in \mathbb{N})$. *Then, with*

$$h_n \sim n^{-\frac{1}{2\beta+d}}$$

for the estimate (4.19) one has

$$\boldsymbol{E}\left\{\int |\sigma_n^2(x) - \sigma^2(x)|^2 \mu(dx)\right\} = O\left(\max\left\{n^{-4\alpha/d}, n^{-2\beta/(2\beta+d)}\right\}\right).$$

(Rate of convergence of the partitioning estimator of the local variance)
With the notation $\sigma^2 \in \Lambda^\beta(D)$ (Lipschitz class) for (4.33), by Theorem 3.2 [11]

$$\liminf_n \inf_{\widetilde{\sigma_n^2}} \sup_{\substack{X \text{ bounded, } (4.32) \\ \sigma^2 \in \Lambda^\beta(D)}} \boldsymbol{E}\left\{\int |\widetilde{\sigma_n^2}(x) - \sigma^2(x)|^2 \mu(dx)\right\} / n^{-\frac{2\beta}{2\beta+d}} \ge C^*$$

(arbitrary fixed τ, β, D, bound of X) holds for some constant $C^* > 0$, where the first infimum is taken over all estimates $\widetilde{\sigma_n^2}$ of σ^2 (lower minimax rate $n^{-2\beta/(2\beta+d)}$ according to Definition 3.1 [11]). For $\beta \le 1$, the assertion in Theorem 4.5 in the case $\alpha \ge \beta d/2(2\beta + d)$ means

$$\limsup_n \sup_{\substack{X \text{ bounded, } (4.32) \\ \sigma^2 \in \Lambda^\beta(D)}} \boldsymbol{E}\left\{\int |\sigma_n^2(x) - \sigma^2(x)|^2 \mu(dx)\right\} / n^{-\frac{2\beta}{2\beta+d}} \le C^{**}$$

(arbitrary fixed τ, β, D, D^* bound of X) for some constant $C^{**} < \infty$. Thus, according to Definition 3.2 in [11], $n^{-\frac{2\beta}{2\beta+d}}$ is an optimal rate attained by σ_n^2, in the case of known m also attained by classical partitioning estimation according to Theorem 4.3 with proof [11].

Proof of Theorem 4.5. Denote by l_n the number of cubes of the partition $\mathcal{P}_n$ that covers the bounded support of μ. It holds $l_n = O(h_n^{-d})$. c_1, $c_2, \ldots$ will be suitable constants. Set

$$W_{n,i} := (Y_i - Y_{N[i,1]})(Y_i - Y_{N[i,2]}),$$

and

$$\widehat{W}_{n,i} := \boldsymbol{E}\{(Y_i - Y_{N[i,1]})(Y_i - Y_{N[i,2]})|X_1, \ldots, X_n\},$$

$$\widehat{\sigma_n^2}(x) := \boldsymbol{E}\{\sigma_n^2(x)|X_1, \ldots, X_n\}$$

$$= \frac{\sum_{i=1}^n \widehat{W}_{n,i} 1_{A_n(x)}(X_i)}{\sum_{i=1}^n 1_{A_n(x)}(X_i)}.$$

In the first step we show

$$\begin{aligned} &\boldsymbol{E}\left\{\int |\sigma_n^2(x) - \sigma^2(x)|^2 \mu(dx)\right\} \\ &\le c_1 n^{-1} h_n^{-d} + \boldsymbol{E}\left\{\int |\widehat{\sigma^2}_n(x) - \sigma^2(x)|^2 \mu(dx)\right\}. \end{aligned} \tag{4.35}$$

We notice

$$\begin{aligned} &\boldsymbol{Var}(W_{n,i}|X_1,\ldots,X_n) \\ &\le \boldsymbol{E}\{W_{n,i}^2|X_1,\ldots,X_n\} \\ &= \boldsymbol{E}\{W_{n,1}^2|X_1,\ldots,X_n\} \\ &\le \frac{1}{2}\boldsymbol{E}\{(Y_1 - Y_{N[1,1]})^4|X_1,\ldots,X_n\} + \frac{1}{2}\boldsymbol{E}\{(Y_1 - Y_{N[1,2]})^4|X_1,\ldots,X_n\} \\ &\le 8\boldsymbol{E}\{Y_1^4|X_1\} + 4\boldsymbol{E}\{Y_{N[1,1]}^4|X_1,\ldots,X_n\} + 4\boldsymbol{E}\{Y_{N[1,2]}^4|X_1,\ldots,X_n\} \\ &\le 8\tau^4 + 4\boldsymbol{E}\left\{\left(\sum_{j=2}^n Y_j 1_{\{N[1,1]=j\}}\right)^4 \middle| X_1,\ldots,X_n\right\} \\ &\quad +4\boldsymbol{E}\left\{\left(\sum_{j=2}^n Y_j 1_{\{N[1,2]=j\}}\right)^4 \middle| X_1,\ldots,X_n\right\} \\ &= 8\tau^4 + 4\sum_{j=2}^n \boldsymbol{E}\{Y_j^4|X_j\}1_{\{N[1,1]=j\}} + 4\sum_{j=2}^n \boldsymbol{E}\{(Y_j^4|X_j\}1_{\{N[1,2]=j\}} \\ &\le 16\tau^4. \end{aligned} \tag{4.36}$$

Analogously to [11], p. 64, we have

$$\begin{aligned} &\boldsymbol{E}\{(\sigma_n^2(x) - \sigma^2(x))^2|X_1,\ldots,X_n\} \\ &= \boldsymbol{E}\{(\sigma_n^2(x) - \widehat{\sigma_n^2}(x))^2|X_1,\ldots,X_n\} + (\widehat{\sigma_n^2}(x) - \sigma^2(x))^2, \end{aligned}$$

where

$$\boldsymbol{E}\{(\sigma_n^2(x) - \widehat{\sigma_n^2}(x))^2|X_1,\ldots,X_n\}$$

$$= \boldsymbol{E}\left\{\left(\frac{\sum_{i=1}^n (W_{n,i} - \widehat{W}_{n,i}) 1_{A_n(x)}(X_i)}{\sum_{i=1}^n 1_{A_n(x)}(X_i)}\right)^2 \middle| X_1, \ldots, X_n\right\}$$

$$= \frac{\sum_{i=1}^n \boldsymbol{Var}(W_{n,i}|X_1, \ldots, X_n) 1_{A_n(x)}(X_i)}{\left(\sum_{i=1}^n 1_{A_n(x)}(X_i)\right)^2}$$

$$\leq \frac{16\tau^4}{\sum_{i=1}^n 1_{A_n(x)}(X_i)} 1_{\{\sum_{i=1}^n 1_{A_n(x)}(X_i) > 0\}}$$

by (4.36). Thus

$$\boldsymbol{E}\left\{\int |\sigma_n^2(x) - \sigma^2(x)|^2 \mu(dx)\right\}$$

$$\leq 16\tau^4 \int \boldsymbol{E}\left\{\frac{1}{\sum_{i=1}^n 1_{A_n(x)}(X_i)} 1_{\{\sum_{i=1}^n 1_{A_n(x)}(X_i) > 0\}}\right\} \mu(dx)$$

$$+ \boldsymbol{E}\left\{\int |\widehat{\sigma_n^2}(x) - \sigma^2(x)|^2 \mu(dx)\right\}$$

$$\leq 32\tau^4 \int \frac{1}{n\mu(A_n(x))} \mu(dx) + \boldsymbol{E}\left\{\int |\widehat{\sigma_n^2}(x) - \sigma^2(x)|^2 \mu(dx)\right\}$$

by Lemma 4.1 [11]

$$\leq 32\tau^4 n^{-1} l_n + \boldsymbol{E}\left\{\int |\widehat{\sigma_n^2}(x) - \sigma^2(x)|^2 \mu(dx)\right\}$$

$$\leq c_1 n^{-1} h_n^{-d} + \boldsymbol{E}\left\{\int |\widehat{\sigma_n^2}(x) - \sigma^2(x)|^2 \mu(dx)\right\},$$

i.e., (4.35) holds.
In the second step we obtain

$$\boldsymbol{E}\left\{\int \left|\widehat{\sigma_n^2}(x) - \frac{\sum_{i=1}^n \widehat{W}_{n,i} 1_{A_n(x)}(X_i)}{n\mu(A_n(x))}\right|^2 \mu(dx)\right\} \leq c_2 n^{-1} h_n^{-d}. \tag{4.37}$$

For, by (4.36),

$$|\widehat{W}_{n,i}| \leq (\boldsymbol{E}\{W_{n,i}^2 | X_1, \ldots, X_n\})^{1/2} \leq 4\tau^2,$$

and, according to p. 465 [11], we have

$$\left|\frac{\sum_{i=1}^n \widehat{W}_{n,i} 1_{A_n(x)}(X_i)}{\sum_{i=1}^n 1_{A_n(x)}(X_i)} - \frac{\sum_{i=1}^n \widehat{W}_{n,i} 1_{A_n(x)}(X_i)}{n\mu(A_n(x))}\right|^2$$

$$\leq 16\tau^4 \left| \frac{\sum_{i=1}^n 1_{A_n(x)}(X_i)}{n\mu(A_n(x))} - 1 \right|^2,$$

further

$$\begin{aligned}
& \boldsymbol{E} \int \left| \frac{\sum_{i=1}^n 1_{A_n(x)}(X_i) - n\mu(A_n(x))}{n\mu(A_n(x))} \right|^2 \mu(dx) \\
& \leq \int \frac{\boldsymbol{Var}(\sum_{i=1}^n 1_{A_n(x)}(X_i))}{n^2\mu(A_n(x))^2} \mu(dx) \\
& \leq \frac{1}{n} \int \frac{1}{\mu(A_n(x))} \mu(dx) \\
& \leq l_n/n \\
& \leq c_2 n^{-1} h_n^{-d}.
\end{aligned}$$

In the third step we show

$$\int \boldsymbol{E} \left\{ \left| \frac{\sum_{i=1}^n \left[\widehat{W}_{n,i} 1_{A_n(x)}(X_i) - \boldsymbol{E}\{\widehat{W}_{n,i} 1_{A_n(x)}(X_i)\} \right]}{n\mu(A_n(x))} \right|^2 \right\} \mu(dx) \leq c_3 n^{-1} h_n^{-d}. \tag{4.38}$$

The left-hand side is bounded by

$$\int \frac{\boldsymbol{Var}\{\sum_{i=1}^n \widehat{W}_{n,i} 1_{A_n(x)}(X_i)\}}{n^2\mu(A_n(x))^2} \mu(dx).$$

As in the proof of Theorem 4.2 we apply the Efron-Stein inequality and obtain, compare (4.18) and notice (4.36),

$$\boldsymbol{Var} \left\{ \sum_{i=1}^n \widehat{W}_{n,i} 1_{A_n(x)}(X_i) \right\} \leq c_4 n \tau^4 \boldsymbol{E}\{1_{A_n(x)}(X)\} = c_5 n\mu(A_n(x)).$$

Further

$$\int \frac{n\mu(A_n(x))}{n^2\mu(A_n(x))^2} \mu(dx) \leq \frac{1}{n} \int \frac{1}{\mu(A_n(x))} \mu(dx) \leq l_n/n \leq c_2 n^{-1} h_n^{-d}.$$

Thus (4.38) is obtained.
In the fourth step we show

$$\int \left| \frac{\sum_{i=1}^{n} \boldsymbol{E}\{\widehat{W}_{n,i} 1_{A_n(x)}(X_i)\}}{n\mu(A_n(x))} - \sigma^2(x) \right|^2 \mu(dx) \leq c_6 \left(h_n^2 + n^{-2/(d+2)} \right), \tag{4.39}$$

i.e.

$$\int \left| \frac{\boldsymbol{E}\{\widehat{W}_{n,1} 1_{A_n(x)}(X_1)\}}{\mu(A_n(x))} - \sigma^2(x) \right|^2 \mu(dx) \leq c_6 \left(h_n^2 + n^{-2/(d+2)} \right). \tag{4.40}$$

According to [19], proof of Theorem 3, or [20], Appendix, we have

$$\begin{aligned} &\boldsymbol{E}\{\widehat{W}_{n,1} 1_{A_n(x)}(X_1)\} \\ &= \boldsymbol{E}\{(Y_1 - m(X_1))^2 1_{A_n(x)}(X_1)\} \\ &\quad + \boldsymbol{E}\left\{ \left(m(X_1) - m(X_{N[1,1]})\right) \left(m(X_1) - m(X_{N[1,2]})\right) 1_{A_n(x)}(X_1) \right\}. \end{aligned}$$

Further

$$\begin{aligned} &\int \left| \frac{\boldsymbol{E}\{(Y_1 - m(X_1))^2 1_{A_n(x)}(X_1)\}}{\mu(A_n(x))} - \sigma^2(x) \right|^2 \mu(dx) \\ &= \int \left| \frac{\int \sigma^2(t) 1_{A_n(x)}(t)\mu(dt)}{\mu(A_n(x))} - \sigma^2(x) \right|^2 \mu(dx) \\ &\leq \int \frac{\left(\int [\sigma^2(t) - \sigma^2(x)] 1_{A_n(x)}(t)\mu(dt)\right)^2}{\mu(A_n(x))} \mu(dx) \\ &\leq D^2 \int \frac{\left(\int \|t - x\| 1_{A_n(x)}(t)\mu(dt)\right)^2}{\mu(A_n(x))^2} \mu(dx) \\ &\quad \text{(by (4.33))} \\ &\leq D^2 d^\beta h_n^{2\beta} \int \frac{\left(\int 1_{A_n(x)}(t)\mu(dt)\right)^2}{\mu(A_n(x))^2} \mu(dx) \leq c_7 h_n^{2\beta}, \end{aligned}$$

and

$$\begin{aligned} &\int \left| \frac{\boldsymbol{E}\left\{ \left(m(X_1) - m(X_{N[1,1]})\right) \left(m(X_1) - m(X_{N[1,2]})\right) 1_{A_n(x)}(X_1) \right\}}{\mu(A_n(x))} \right|^2 \mu(dx) \\ &\leq \frac{1}{4} \int \frac{\left(\boldsymbol{E}\left\{ \left|m(X_1) - m(X_{N[1,1]})\right|^2 1_{A_n(x)}(X_1) \right\} \right)^2}{\mu(A_n(x))^2} \mu(dx) \\ &\quad + \frac{1}{4} \int \frac{\left(\boldsymbol{E}\left\{ \left|m(X_1) - m(X_{N[1,2]})\right|^2 1_{A_n(x)}(X_1) \right\} \right)^2}{\mu(A_n(x))^2} \mu(dx) \end{aligned}$$

(by the Cauchy-Schwarz inequality)

$$\leq \frac{1}{4}\int \frac{\boldsymbol{E}\left\{\left[\left|m(X_1)-m(X_{N[1,1]})\right|^4+\left|m(X_1)-m(X_{N[1,2]})\right|^4\right]1_{A_n(x)}(X_1)\right\}}{\mu(A_n(x))}\mu(dx)$$

$$\leq \frac{1}{4}D^{*4}\left[\boldsymbol{E}\left\{\|X_{N[1,1]}-X_1\|^{4\alpha}\right\}+\boldsymbol{E}\left\{\|X_{N[1,2]}-X_1\|^{4\alpha}\right\}\right]$$

(by (4.34) and $\int \left[1_{A_n(x)}(t)/\mu(A_n(x))\right]\mu(dx)\leq 1$ for each $t\in\mathbb{R}^d$)

$$\leq c_8 n^{-4\alpha/d},$$

where the last inequality immediately follows from Theorem 3 in Liitiäinen et al. (2010). Thus (4.40) and (4.39) are obtained.

In the last step we gather (4.35), (4.37), (4.38), (4.39) and obtain

$$\begin{aligned}
&\boldsymbol{E}\left\{\int |\sigma_n^2(x)-\sigma^2(x)|^2\mu(dx)\right\}\\
&\leq c_9\left(n^{-1}h_n^{-d}+h_n^{2\beta}+n^{-4\alpha/d}\right)\\
&\leq c_{10}\max\left\{n^{-4\alpha/d},n^{-2\beta/(2\beta+d)}\right\}
\end{aligned}$$

by the choice of (h_n). Thus the assertion is obtained. □

Chapter 5
Local Variance Estimation for Censored Observations

5.1 Introduction

This chapter deals with local variance estimation in the presence of randomly censored data.
Let (X, Y, C), (X_1, Y_1, C_1), $(X_2, Y_2, C_2), \dots$ i.i.d. $\mathbb{R}^d \times \mathbb{R}_+ \times \mathbb{R}_+$-valued random vectors. X is the random vector of covariates with distribution μ, which, e.g., in medical applications contains information about a human taking part in a medical study around an illness. Y represents the survival time of the patient. C represents the censoring time. Moreover, we introduce the variable T, defined as minimum of Y and C, and the variable δ, containing the information whether there is or not censoring. This yields a set of data

$$\{(X_1, T_1, \delta_1), \dots, (X_n, T_n, \delta_n)\},$$

with

$$\begin{cases} \delta_i = 1 \text{ for } Y_i \leq C_i \\ \delta_i = 0 \text{ for } Y_i > C_i, \end{cases}$$

and

$$T_i = \min\{Y_i, C_i\},$$

for $i = 1, \dots, n$. In medical studies the observation of the survival time of the patient is sometimes incomplete due to RIGHT censoring formulated just before. It could, for example, happen that the patient is alive at the termination of a medical study, or that he dies by other causes than those under study, or, trivially, that the patient moves and the hospital loses information about him. For more details see for example [11], Chapter 26.
We introduce now the so-called survival functions

$$F(t) = \boldsymbol{P}(Y > t),$$

$$G(t) = \boldsymbol{P}(C > t),$$

and

$$K(t) = \boldsymbol{P}(T > t) = F(t)G(t).$$

Introduce also

$$F^*(t) := \boldsymbol{P}(Y^2 > t) = F(\sqrt{t}),$$

$$K^*(t) := \boldsymbol{P}(T^* > t) = F^*(t)G(t) = F(\sqrt{t})G(t),$$

where $T^* = \min\{Y^2, C\}$.
The survival functions map the event of survival onto time and are therefore monotone decreasing. Define

$$T_F := \sup\{y : \ F(y) > 0\},$$

$$T_G := \sup\{y : \ G(y) > 0\},$$

$$T_K := \sup\{y : \ K(y) > 0\} = \min\{T_F, T_G\},$$

and notice that

$$T_{F^*} := \sup\{y : \ F^*(y) > 0\} = T_F$$

and

$$T_{K^*} := \sup\{y : \ K^*(y) > 0\} = \min\{T_{F^*}, T_G\} = \min\{T_F, T_G\} = T_K.$$

For our intents we require the following conditions, which are required throughout the whole chapter 5:

(A1) C and (X, Y) are independent,
(A2) $\exists L > 0$, such that $\boldsymbol{P}\{\max\{Y, Y^2\} \leq L\} = 1$ and $\boldsymbol{P}\{C > L\} > 0$.
G is continuous.
(A3) $\forall 0 < T'_K < T_K : \ \boldsymbol{P}\{0 \leq Y \leq T'_K\} < 1$, $\boldsymbol{P}\{0 \leq Y^2 \leq T'_K\} < 1$
F is continuous in a neighborhood of T_K and in a neighborhood of $\sqrt{T_K}$.

As we already said, under censoring the information about the survival time of a patient are incomplete in the sense that sometimes we cannot observe Y_i but only C_i with the indication that it is not the real life time (by δ_i). Therefore the random triple (X, T, δ) does not identify anymore the conditional distribution of Y given X. To achieve it, we need an additional assumption, that is **(A1)**: the censoring time C is independent of the common distribution of the survival time Y and the patient data X. In the medical applications **(A1)** is fulfilled in the case the censoring takes place

regardless the characteristics of the patients and depends only on external factors not related to the information represented by the covariate X. Examples of this situation are the (random) termination of a study, which does not depend on the person who participated to it or the interruption of the cooperation of the patient to the medical study, maybe because of luck of enthusiasm.
The first part of **(A2)** is obviously fulfilled because of the intrinsic boundedness of Y (survival time of a human being!). The second part of **(A2)**, the positivity of $\boldsymbol{P}\{C > L\}$ means that not the whole censoring process takes place in $[0, L]$. In practice, it means that there is the possibility to extend the medical study, so that, with positive probability, C is larger than the bound L of Y.
The continuity of G will be necessary for the convergence of the estimator G_n of G, that we introduce in following. Moreover, for this estimator the assumption **(A3)** allows giving a rate of convergence on the whole interval $[0, T_K]$.
For unknown F and G, Kaplan and Meier [15] proposed two estimates, F_n and G_n, respectively, the product-limit estimates (see for example [11], pp. 541, 542). In medical research, the Kaplan-Meier estimate is used to measure the fraction of patients living for a certain amount of time after treatment. Also in economics it is common, for measuring the length of time people remain unemployed after a job loss. In engineering, it can be used to measure the time until failure of machine parts.
Let F_n and G_n be the Kaplan-Meier estimates of F and G, respectively, which are defined as

$$F_n(t) = \begin{cases} \prod_{i=1,\ldots,n\ T(i)\leq t} \left(\frac{n-i}{n-i+1}\right)^{\delta_{(i)}} & t \leq T(n) \\ 0 & \text{otherwise} \end{cases}$$

and

$$G_n(t) = \begin{cases} \prod_{i=1,\ldots,n\ T(i)\leq t} \left(\frac{n-i}{n-i+1}\right)^{1-\delta_{(i)}} & t \leq T(n) \\ 0 & \text{otherwise,} \end{cases}$$

where $((T(1), \delta(1)), \ldots, (T(n), \delta(n)))$ are the n pairs of observed (T_i, δ_i) set in increasing order.
Fan and Gijbels [8] introduced a transformation $\widetilde{Y}$ of the variable T with

$$\widetilde{Y} := \frac{\delta T}{G(T)},$$

and correspondingly:

$$\widetilde{Y}_i = \frac{\delta_i T_i}{G(T_i)},$$

under known survival function G, and finally

$$\widetilde{Y}_{n,i} = \frac{\delta_i T_i}{G_n(T_i)},$$

where G is estimated by Kaplan-Meier estimator G_n in the case it is unknown.
Define then

$$\widetilde{Y^2} := \frac{\delta T^2}{G(T)} \tag{5.1}$$

and their observations (G is known)

$$\widetilde{Y_i^2} = \frac{\delta_i T_i^2}{G(T_i)}, \tag{5.2}$$

and, for unknown G,

$$\widetilde{Y_{n,i}^2} = \frac{\delta_i T_i^2}{G_n(T_i)}. \tag{5.3}$$

Notice that $\widetilde{Y^2} \neq \widetilde{Y}^2 = \left(\frac{\delta T}{G(T)}\right)^2$.
The first part of assumption **(A2)** is equivalent to $0 \leq Y \leq L$, $Y^2 \leq L$ a.s., and it imply $T_K \leq L$ a.s.
Because of $0 \leq T_i \leq T_K \leq L$ for $i = 1, \ldots, n$ with $G(L) = \boldsymbol{P}\{C > L\} > 0$ we get

$$1 \geq G(T_{(1)}) \geq \cdots \geq G(T_{(n)}) \geq G(T_K) \geq G(L) > 0 \quad a.s. \tag{5.4}$$

For fixed n also G_n is monotone decreasing

$$1 \geq G_n(T_{(1)}) \geq \cdots \geq G_n(T_{(n)}) \geq G_n(T_K) \geq G_n(L) > 0 \quad a.s. \tag{5.5}$$

Therefore, because of the boundedness of Y from (5.5) and the convergence theorem of Stute and Wang [33] follows

$$\widetilde{Y}_{n,i} < U < \infty \text{ and } \widetilde{Y_{n,i}^2} < U < \infty \quad a.s. \tag{5.6}$$

(5.6) follows from (5.5) and $G_n(L) \to G(L)$ a.s. (the latter because of [11], Theorem 26.1)
For the transformation $\widetilde{Y}$ and $\widetilde{Y^2}$ the following nice properties can be shown:

$$
\begin{aligned}
&\boldsymbol{E}\left\{\widetilde{Y}|X\right\} \\
&= \boldsymbol{E}\left\{\left.\frac{1_{\{Y<C\}}\min\{Y,C\}}{G(\min\{Y,C\})}\right|X\right\} \\
&= \boldsymbol{E}\left\{\boldsymbol{E}\left\{\left.1_{\{Y<C\}}\frac{Y}{G(Y)}\right|X,Y\right\}\middle|X\right\} \\
&= \boldsymbol{E}\left\{\frac{Y}{G(Y)}\underbrace{\boldsymbol{E}\{1_{\{Y<C\}}|X,Y\}}_{=G(Y)\text{ by }\mathbf{(A1)}}|X\right\} \\
&= \boldsymbol{E}\{Y|X\}
\end{aligned}
\tag{5.7}
$$

and

$$
\begin{aligned}
&\boldsymbol{E}\left\{\widetilde{Y^2}|X\right\} \\
&= \boldsymbol{E}\left\{\left.\frac{1_{\{Y<C\}}\min\{Y^2,C\}}{G(\min\{Y,C\})}\right|X\right\} \\
&= \boldsymbol{E}\left\{\boldsymbol{E}\left\{\left.1_{\{Y<C\}}\frac{Y^2}{G(Y)}\right|X,Y\right\}\middle|X\right\} \\
&= \boldsymbol{E}\left\{\frac{Y^2}{G(Y)}\underbrace{\boldsymbol{E}\{1_{\{Y<C\}}|X,Y\}}_{=G(Y)\text{ by }\mathbf{(A1)}}|X\right\} \\
&= \boldsymbol{E}\{Y^2|X\}.
\end{aligned}
\tag{5.8}
$$

(5.7) and (5.8) mean that the conditional expectation of the transformed censored variable with respect to X equals the conditional expectation of the uncensored variable with respect to X (under **(A1)**). This implies that under known G, in the case that only the pair (T_i, δ_i) instead of (Y_i) is available,

$$\frac{1}{n}\sum_{i=1}^{n}\frac{\delta_i T_i}{G(T_i)}$$

is an unbiased estimate of $\boldsymbol{E}\{Y\}$.
Observe now, that

$$\boldsymbol{Var}\{\widetilde{Y}|X\} = \boldsymbol{Var}\left\{\frac{\delta T}{G(T)}|X\right\} = \boldsymbol{E}\left\{\left.\left(\frac{\delta T}{G(T)}\right)^2\right|X\right\}$$

$$- \, \boldsymbol{E}^2\left\{\frac{\delta T}{G(T)}\middle| X\right\} = \boldsymbol{E}\left\{\frac{(\delta T)^2}{G^2(T)}\middle| X\right\} - \boldsymbol{E}^2\{Y|X\}$$

$$\overset{\delta^2=\delta}{\underset{\downarrow}{=}} \boldsymbol{E}\left\{\frac{1_{\{Y\leq C\}}(\min\{Y,C\})^2}{G^2(\min\{Y,C\})}\middle| X\right\} - m^2(X)$$

$$= \boldsymbol{E}\left\{\boldsymbol{E}\left\{1_{\{Y\leq C\}}\frac{Y^2}{G^2(Y)}\middle| X,Y\right\}\middle| X\right\} - m^2(X)$$

$$= \boldsymbol{E}\left\{\frac{Y}{G^2(Y)}\underbrace{\boldsymbol{E}\{1_{\{Y<C\}}|X,Y\}}_{=G(Y)\text{ by }\mathbf{(A1)}}|X\right\} - m^2(X)$$

$$= \boldsymbol{E}\left\{\frac{Y^2}{G(Y)}\middle| X\right\} - m^2(X).$$

The aim of this chapter is to present estimators (σ_n^2) of the local variance with various estimation approaches (least squares via plug-in in Section 5.2, local averaging via plug-in in Section 5.3, and partitioning via nearest neighbors in Section 5.4) in the censored case, under the definitions and assumptions of this Section (5.1).

5.2 Censored Least Squares Estimation via Plug-In

This section deals with the least squares estimate of the local variance function under censored data. For the least squares estimate under uncensored data see again chapter 2.
Recall now the definitions (5.1), (5.2) and (5.3) and the assumptions **(A1)**-**(A3)** of Section 5.1.
Moreover, without loss of generality, we assume in this section that X is bounded.
Introduce now the following variables

$$\widetilde{Z} := \widetilde{Y^2} - \widetilde{m}^2(X) = \widetilde{Y^2} - m^2(X)$$

and

$$\widetilde{Z}_i := \widetilde{Y_i^2} - \widetilde{m}^2(X_i) = \widetilde{Y_i^2} - m^2(X_i),$$

under known m and known survival function G, where the equalities hold because of (5.7).

However m and G are typically unknown, when the distribution of (X_i, T_i, δ_i) is unknown. In this case, define

$$\widetilde{Z}_{n,i} := \widetilde{Y^2_{n,i}} - \widetilde{m}^2_n(X_i)^{(LS)}$$

with

$$\widetilde{m}_n(\cdot)^{(LS)} := \arg\min_{f \in \mathcal{F}_n} \frac{1}{n} \sum_{i=1}^{n} |f(X_i) - \widetilde{Y}_i|^2, \tag{5.9}$$

where $f : \mathbb{R}^d \to \mathbb{R} \in \mathcal{F}_n$, $\mathcal{F}_n$ being a suitable function space.
$\widetilde{m}_n(\cdot)^{(LS)}$ is a least squares estimator of the regression function in the censored case. [16] and [22] introduced different least squares estimates of the regression function under censoring.
Recall now the definition of the local variance

$$\sigma^2(x) := \boldsymbol{E}\{(Y - m(X))^2 | X = x\}$$

and the definition $Z = Y^2 - m^2(X)$, and notice that the local variance function is a regression on (X, Z).
We can therefore define the least squares estimator of the local variance, analogously to (5.9) as

$$\widetilde{\sigma}^2_n(\cdot)^{(LS)} := \arg\min_{g \in \mathcal{G}_n} \frac{1}{n} \sum_{i=1}^{n} |g(X_i) - \widetilde{Z}_{n,i}|^2, \tag{5.10}$$

where $g : \mathbb{R}^d \to \mathbb{R} \in \mathcal{G}_n$, $\mathcal{G}_n$ being a suitable function space.
The following theorem shows consistency of this estimator. It is analogous to Theorem 2.1 for the uncensored case.

Theorem 5.1. *Assumptions* ***(A1)-(A2)*** *hold. Let $\mathcal{G}_n$ be defined as a subset of a linear space, consisting of nonnegative real-valued functions on $\mathbb{R}^d$ bounded by L^*, with dimension $D_n \in \mathbb{N}$, with the properties $\mathcal{G}_n \uparrow$, $D_n \to \infty$ for $n \to \infty$ but $\frac{D_n}{n} \to 0$. Furthermore $\cup_n \mathcal{G}_n$ is required to be dense in the subspace of $L_2(\mu)$ consisting of the nonnegative functions in $L_2(\mu)$ bounded by L^*. Let also $\mathcal{F}_n$ be defined as a subset of a linear space of real-valued functions on $\mathbb{R}^d$ absolutely bounded by L, with dimension $D'_n \in \mathbb{N}$, with the properties $\mathcal{F}_n \uparrow$, $D'_n \to \infty$ for $n \to \infty$ but $\frac{D'_n}{n} \to 0$. Furthermore $\cup_n \mathcal{F}_n$ is required to be a dense subset of $C^0_{0,L}(\mathbb{R}^d)$ (with respect to the max norm), where $C^0_{0,L}(\mathbb{R}^d)$ denotes the space of continuous real valued functions on $\mathbb{R}^d$ absolutely bounded by L, with compact support.*

Then

$$\int |\widetilde{\sigma}_n^2(x)^{(LS)} - \sigma^2(x)|^2 \mu(dx) \xrightarrow{P} 0. \tag{5.11}$$

(Consistency of the least squares estimator of the local variance under censoring)

Proof of Theorem 5.1. Because of the assumption **(A2)** from $\boldsymbol{P}\{0 \leq Y^2 \leq L\} = 1$ *a.s.* follows $0 \leq m^2(x) \leq L < \infty$ *a.s.*, $\forall x \in [0,1]^d$. From this and by (5.1) and (5.4) follows that $|\widetilde{Y^2} - m^2(X)|$ is bounded, say, smaller than $\beta > 0$ (and therefore sub-Gaussian, cf. Theorem 2.1).
With this, as in the proof of Theorem 2.1 we obtain that there exists a generic positive constant c depending only from L and β with the following property:

$$\boldsymbol{P}\Bigg\{\int |\widetilde{\sigma}_n^2(x)^{(LS)} - \sigma^2(x)|^2 \mu(dx) > c \cdot \left(\frac{1}{n}\sum_{i=1}^{n} |Z_i - \widetilde{Z}_{n,i}|^2 + \frac{D_n}{n} + \inf_{g \in \mathcal{G}_n} \int |g(x) - \sigma^2(x)|^2 \mu(dx)\right)\Bigg\} \to 0. \tag{5.12}$$

We notice

$$\frac{1}{n}\sum_{i=1}^{n} |Z_i - \widetilde{Z}_{n,i}|^2 \leq \frac{2}{n}\sum_{i=1}^{n} |\widetilde{m}_n^2(X_i)^{(LS)} - m^2(X_i)|^2 + \frac{2}{n}\sum_{i=1}^{n} |\widetilde{Y_i^2} - \widetilde{Y_{n,i}^2}|^2$$
$$\leq \frac{2}{n}\sum_{i=1}^{n} |\widetilde{m}_n(X_i)^{(LS)} - m(X_i)|^2 + \frac{2}{n}\sum_{i=1}^{n} |\widetilde{Y_i^2} - \widetilde{Y_{n,i}^2}|^2, \tag{5.13}$$

because of the boundedness of m and the structure of m_n.
But

$$\frac{1}{n}\sum_{i=1}^{n} \left|\widetilde{Y_i^2} - \widetilde{Y_{n,i}^2}\right|^2 = \frac{1}{n}\sum_{i=1}^{n} \left|\widetilde{Y}_i + \widetilde{Y}_{n,i}\right|^2 \cdot \left|\widetilde{Y}_i - \widetilde{Y}_{n,i}\right|^2$$
(because of (5.6))
$$\leq \frac{1}{n} 4U^2 \sum_{i=1}^{n} \left|\widetilde{Y}_i - \widetilde{Y}_{n,i}\right|^2 \to 0, \tag{5.14}$$

almost surely and therefore in probability, due to Lemma 2 in [22], p. 52, under **(A2)**, having set there $\alpha = -1$.
Now, it remains to prove

$$\frac{1}{n}\sum_{i=1}^{n}|\widetilde{m}_n(X_i)^{(LS)} - m(X_i)|^2 \xrightarrow{P} 0.$$

Via conditioning with respect to $(X_1, \ldots, X_n)$, by [16], Lemma 3, we obtain, for a constant c depending only on L and β' where $|\widetilde{Y} - m(X)| \leq \beta'$

$$\boldsymbol{P}\left\{\frac{1}{n}\sum_{i=1}^{n}|\widetilde{m}_n(X_i)^{(LS)} - m(X_i)|^2 \right.$$
$$\left. > c\left(\frac{1}{n}\sum_{i=1}^{n}|Y_i - \widetilde{Y}_{n,i}|^2 + \frac{D_n'}{n} + \min_{f\in\mathcal{F}_n}\frac{1}{n}\sum_{i=1}^{n}|f(X_i) - m(X_i)|^2\right)\right\} \to 0 \tag{5.15}$$

where $\frac{D_n'}{n} \to 0$ and $\frac{1}{n}\sum_{i=1}^{n}|Y_i - \widetilde{Y}_{n,i}|^2 \xrightarrow{P} 0$ (see (5.14)).
Regarding the last term in (5.15), we choose for an arbitrary $\varepsilon' > 0$ a continuous function with compact support $\hat{m}$ such that $\boldsymbol{E}|\hat{m}(X) - m(X)|^2 \leq \varepsilon'$.
We observe

$$\min_{f\in\mathcal{F}_n}\frac{1}{n}\sum_{i=1}^{n}|f(X_i) - m(X_i)|^2 \leq$$
$$\leq 2\underbrace{\min_{f\in\mathcal{F}_n}\frac{1}{n}\sum_{i=1}^{n}|f(X_i) - \hat{m}(X_i)|^2}_{\to 0 \ a.s.} \tag{5.16}$$
$$+\, 2 \underbrace{\frac{1}{n}\sum_{i=1}^{n}|\hat{m}(X_i) - m(X_i)|^2}_{a.s.\ \to \boldsymbol{E}\{|\hat{m}(X)-m(X)|^2\}\leq\varepsilon' \ (\text{ Strong Law of Large Numbers})}$$

where (5.16) follows from
$\min_{f\in\mathcal{F}_n}\frac{1}{n}\sum_{i=1}^{n}|f(X_i) - \hat{m}(X_i)|^2 \leq \min_{f\in\mathcal{F}_n}\left[\sup_x|f(x) - \hat{m}(x)|^2\right] \to 0$.
Since almost sure convergence implies convergence in probability, by Lemma (2.1) the assertion follows. □
We want now to give a rate of convergence for the local variance estimator. The following theorem corresponds to Mathe's ([22], Satz 4) result on rate of convergence of least square regression estimation under censoring. In view of

rate of convergence we consider till now, as Mathe [22] did, $(p, \mathit{\Gamma})$-smooth functions. Later on, in Sections 5.3 and 5.4, we consider only Lipschitz continuos functions, i.e., $p = 1$.
Moreover we conduct a modification of the estimators of the regression function (5.9) and of the local variance (5.10) providing definitions m_n^* and $\sigma_n^2{}^*$, respectively. The aim of this modification is to cut the estimators at height L and L^*, respectively, as Mathe [22] did, so that $0 \leq m_n(x)^* \leq L$ as $0 \leq m(x) \leq L$ and, analogously, $0 \leq \sigma_n^2{}^*(x) \leq L^*$ as $0 \leq \sigma^2(x) \leq L^*$.
Set therefore now

$$\widetilde{m}_n(x)^* := T_{[0,L]}(x)\widetilde{m}_n(x)$$

and

$$\widetilde{\sigma}_n^2(x)^* = T_{[0,L^*]}(x)\widetilde{\sigma}_n^2(x),$$

where

$$T_{[a,b]}(x) = \begin{cases} b & \text{if } x > b, \\ x & \text{if } a \leq x \leq b, \\ a & \text{if } x < a. \end{cases}$$

For $f : \mathbb{R}^d \to R$ the function $T_{[a,b]}f : \mathbb{R}^d \to \mathbb{R}$ is defined as

$$(T_{[a,b]}f)(x) := T_{[a,b]}(f(x)) \quad (x \in \mathbb{R}^d).$$

Recall now the definitions of univariate and multivariate B-splines and of splines space from Chapter 2: Definitions 2.1, 2.2 and 2.3. We choose as suitable function space for the minimization problem in (5.10) the space of B-spline functions $S_{K_n,M}([0,1]^d)$, like Mathe did [22].
We set from now on for simplicity $\widetilde{m}_n = \widetilde{m}_n^*$ and $\widetilde{\sigma}_n^2 = \widetilde{\sigma}_n^2{}^*$, with $L = L^*$.
The proofs of Satz 2, Satz 3 and Satz 4 in Mathe [22] show (m $(p, \mathit{\Gamma})$-smooth)

$$\begin{aligned}
&\frac{1}{n}\sum_{i=1}^{n}|\widetilde{m}_n(X_i)^{(LS)} - m(X_i)|^2 \\
&= O_P\left(\frac{1}{n}\sum_{i=1}^{n}|\widetilde{Y}_{n,i} - \widetilde{Y}_i|^2 + \frac{K_n^d \log n}{n}\right. \\
&\quad \left. + \inf_{f \in S_{K_n,M}([0,1]^d)} \frac{1}{n}\sum_{i=1}^{n}|f(X_i) - m(X_i)|^2\right) \\
&= O_P\left(\frac{1}{n}\sum_{i=1}^{n}|\widetilde{Y}_{n,i} - \widetilde{Y}_i|^2 + \mathit{\Gamma}^{\frac{2p}{2p+d}}\left(\frac{\log n}{n}\right)\right)
\end{aligned}$$

$$= O_P\left(\left(\frac{\log n}{n}\right)^{\frac{1}{3}} + \Gamma^{\frac{2p}{2p+d}}\left(\frac{\log n}{n}\right)^{\frac{2p}{2p+d}}\right). \tag{5.17}$$

The following theorem gives the rate of convergence of the least squares estimator of the local variance under censoring.

Theorem 5.2. *Assumptions **(A1)-(A3)** hold.* $X \in [0,1]^d$ *almost surely. Moreover, let* $\Gamma > 0$, $\Lambda > 0$ *and* $p = k + \beta$ *for some* $k \in \mathbb{N}_0$ *and* $\beta \in (0,1]$. m *and* σ^2 *are* (p,Γ) *and* (p,Λ)*-smooth, respectively, that is, for every* $\alpha = (\alpha_1,\ldots,\alpha_d)$, $\alpha_j \in \mathbb{N}_0$, $\sum_{j=1}^d \alpha_j = k$

$$\left|\frac{\partial^k m}{\partial x_1^{\alpha_1},\ldots,\partial x_d^{\alpha_d}}(x) - \frac{\partial^k m}{\partial x_1^{\alpha_1},\ldots,\partial x_d^{\alpha_d}}(z)\right| \le \Gamma\|x-z\|^\beta \quad x,\ z \in \mathbb{R}^d$$

and

$$\left|\frac{\partial^k \sigma^2}{\partial x_1^{\alpha_1},\ldots,\partial x_d^{\alpha_d}}(x) - \frac{\partial^k \sigma^2}{\partial x_1^{\alpha_1},\ldots,\partial x_d^{\alpha_d}}(z)\right| \le \Lambda\|x-z\|^\beta \quad x,\ z \in \mathbb{R}^d$$

($\| \ \|$ denoting the Euclidean norm).
Identify $\mathcal{F}_n$ *and* $\mathcal{G}_n$ *with* $S_{K_n',M}([0,1]^d)$ *and* $S_{K_n,M}([0,1]^d)$*, respectively, with respect to an equidistant partition of* $[0,1]^d$ *into*

$$K_n' = \lceil \Gamma^{\frac{2}{2p+d}}\left(\frac{\log n}{n}\right)^{\frac{1}{2p+d}}\rceil$$

for $\mathcal{F}_n$ *and*

$$K_n = \lceil \Lambda^{\frac{2}{2p+d}}\left(\frac{\log n}{n}\right)^{\frac{1}{2p+d}}\rceil,$$

for $\mathcal{G}_n$*, respectively.*
Then

$$\int \left|\tilde{\sigma}_n^2(x) - \sigma^2(x)\right|^2 \mu(dx) = O_P\left(\left(\frac{\log n}{n}\right)^{\frac{1}{3}} + \left(\frac{\log n}{n}\right)^{\frac{2p}{2p+d}}\right).$$

(Rate of convergence of the least squares estimator of the local variance under censoring)

Proof. We use now (5.12). Because of the dimension $D_n = c \cdot K_n$ of $\mathcal{G}_n$ it follows

$$\frac{D_n}{n} \le O\left(\frac{\log n}{n}\right)^{-\frac{2p}{2p+d}}.$$

From the (p, Λ)-smoothness of σ^2 and the definition of $\mathcal{G}_n$ we can conclude (cf. proof of Satz 3 in [22])

$$\inf_{g\in\mathcal{G}_n} \int |g(x)-\sigma^2(x)|^2\mu(dx) \leq \inf_{g\in\mathcal{G}_n} \sup_{x\in[0,1]^d} |g(x)-\sigma^2(x)|^2$$

$$\leq c\Lambda^2\left(\frac{1}{K_n}\right)^{2p} \leq O\left(\frac{\log n}{n}\right)^{-\frac{2p}{2p+d}} \tag{5.18}$$

for n sufficiently large. In view of the assertion it remains to give a rate for

$$\frac{1}{n}\sum_{i=1}^{n}\left|Z_i-\widetilde{Z}_i\right|^2.$$

Now we use (5.13). It holds

$$\begin{aligned}
&\frac{1}{n}\sum_{i=1}^{n}\left|\widetilde{Y_i^2}-\widetilde{Y_{n,i}^2}\right|^2\\
=\ &\frac{1}{n}\sum_{i=1}^{n}\delta_i T_i^2\left|\frac{1}{G(T_i)}-\frac{1}{G_n(T_i)}\right|^2\\
\overset{T_i\leq T_K\leq L}{\leq}\ &L^4\frac{1}{n}\sum_{i=1}^{n}\sup_{0\leq t\leq T_K}\left|\frac{1}{G(t)}-\frac{1}{G_n(t)}\right|^2\\
\leq\ &L^4\frac{1}{G_n^2(T_K)G^2(T_K)}\frac{1}{n}\sum_{i=1}^{n}\sup_{0\leq t\leq T_K}|G_n(t)-G(t)|^2\\
=\ &O_P\left(\left(\frac{\log n}{n}\right)^{\frac{1}{3}}\right)
\end{aligned} \tag{5.19}$$

(as shown in proof of Satz 4, [22], under **(A3)**). Thus it remains to show

$$\frac{1}{n}\sum_{i=1}^{n}\left|\widetilde{m}_n(X_i)^{(LS)}-m(X_i)\right|^2$$

$$=O_P\left(\left(\frac{\log n}{n}\right)^{\frac{1}{3}}+\Gamma^{\frac{2p}{2p+d}}\left(\frac{\log n}{n}\right)^{\frac{2p}{2p+d}}\right).$$

But it holds also true, due to (5.17) and therefore, the assertion follows. □

Remark 5.1. Notice that for $d > 4p$ the optimal rate is $\left(\frac{\log n}{n}\right)^{\frac{2p}{2p+d}}$:

$$\text{for } p > 0 \qquad \frac{2p}{2p+d} = \frac{1}{1+\frac{d}{2p}} < \frac{1}{3}$$

and therefore

$$\left(\frac{\log n}{n}\right)^{\frac{2p}{2p+d}} > \left(\frac{\log n}{n}\right)^{\frac{1}{3}}.$$

For $d \le 4p$ the rate is $\left(\frac{\log n}{n}\right)^{\frac{1}{3}}$. It implies that for the covariate dimension $d \le 4p$ in the censored situation the convergence of the local variance estimator is slower than in the uncensored case.

5.3 Censored Local Averaging Estimation via Plug-In

The aim of this section is to introduce a local averaging estimator ot the local variance function in the censoring case. The chapter 3 introduces the local averaging estimator in the uncensored case. Now, recall again the definitions and assumptions of section 5.1, in particular assumptions **(A1)**, **(A2)** and **(A3)** hold. One defines a new random variable

$$\widetilde{Z} := \widetilde{Y^2} - m^2(X)$$

(notice that $\widetilde{m}(X) := \boldsymbol{E}\{\widetilde{Y}|X\} = m(X)$ because of (5.7)) and in context of observations in the case of known regression function

$$\widetilde{Z}_i := \widetilde{Y_i^2} - m^2(X_i)$$

and finally in the case of unknown regression function

$$\widetilde{Z}_{n,i} = \widetilde{Y_{n,i}^2} - \widetilde{m}_n^2(X_i)^{(LA)},$$

where

$$\widetilde{m}_n(x)^{(LA)} := \sum_{i=1}^{n} W_{n,i}(x)\widetilde{Y}_{n,i} \tag{5.20}$$

(Notice that usually m is unknown and has to be estimated. In this way one has a plug-in method.) Note that the local variance function is a regression on the pair (X, Z).

This motivates the construction of a family of estimates of the local variance that have the form

$$\widetilde{\sigma}_n^2(x)^{(LA)} := \sum_{i=1}^{n} W_{n,i}(x) \cdot \widetilde{Z}_{n,i}, \tag{5.21}$$

with weights $W_{n,i}(x)$ different from the weights $W_{n,i}(x)$ in (5.20). The weights $W_{n,i}(x)$ can take different forms. In the literature partitioning weights are used, defined by

$$W_{n,i}(x, X_1, \dots, X_n) = \frac{1_{A_n(x)}(X_i)}{\sum_{l=1}^{n} 1_{A_n(x)}(X_l)} \tag{5.22}$$

($A_n(x)$ denoting the $A_{n,j}$ of the partitioning sequence $\{A_{n,j}\}$ containing $x \in \mathbb{R}^d$), with $0/0 := 0$.
Further kernel weights are used, especially with symmetric kernel $K : \mathbb{R}^d \to [0, \infty)$, satisfying $1_{S_{0,R}} \geq K(x) \geq b 1_{S_{0,r}}(x)$ $(0 < r \leq R < \infty,\ b > 0)$, defined by

$$W_{n,i}(x, X_1, \dots, X_n) = \frac{K\left(\frac{x - X_i}{h_n}\right)}{\sum_{l=1}^{n} K\left(\frac{x - X_l}{h_n}\right)}, \tag{5.23}$$

with bandwidth $h_n > 0$ and $0/0 := 0$ again. $S_{0,r}$ denotes the sphere with radius $r > 0$ centered in 0.
Finally, nearest neighbor weights are also frequently used, defined by

$$W_{n,i}(x, X_1, \dots, X_n) = \frac{1}{k_n} 1_{\{X_i \text{ is among the } k_n \text{ NNs of } x \text{ in } \{X_1, \dots, X_n\}\}} \tag{5.24}$$

$(2 \leq k_n \leq n)$, here usually assuming that ties occur with probability 0. This can be obtained for example by use of a suitable additional component of X_i (compare [11], pp. 86, 87).
We distinguish local averaging methods in the auxiliary estimates $\widetilde{m}_n^{(LA)}$ in (5.20) and in the estimates $\widetilde{\sigma}_n^{2\ (LA)}$ in (5.21), indicating the weights by $W_{n,i}^{(2)}$ for the regression function estimator and $W_{n,i}^{(1)}$ for the local variance estimator.

$$W_{n,i}^{(2)}(x) = W_{n,j}^{(2)}(x, X_1, \dots, X_n)$$

are of partitioning type, with partitioning sequence $\left\{A_{n,j}^{(2)}\right\}$, or of kernel type, with kernel $K^{(2)}$ and bandwidths $h_n^{(2)}$, or of nearest neighbor type, with $k_n^{(2)}$ nearest neighbors.

$$W_{n,i}^{(1)}(x) = W_{n,i}^{(1)}(x, X_1, \ldots, X_n)$$

are of partitioning type, with partitioning sequence $\left\{A_{n,j}^{(1)}\right\}$, or of kernel type, with kernel $K^{(1)}$ and bandwidths $h_n^{(1)}$. (Nearest neighbor weights will not be used for $W_{n,i}^{(1)}(x)$.)
The following theorem shows consistency of the local variance estimator (5.21).

Theorem 5.3. *Let the assumptions* ***(A1)*** *and* ***(A2)*** *hold. For partitioning weights defined according to (5.22) assume that, for each sphere S centered at the origin*

$$\lim_{n\to\infty} \max_{A_{n,j}^{(l)}\cap S\neq\emptyset} \operatorname{diam}(A_{n,j}^{(l)}) = 0, \quad l = 1,2, \tag{5.25}$$

$$\lim_{n\to\infty} \frac{|\{j : A_{n,j}^{(l)} \cap S \neq \emptyset\}|}{n} = 0, \quad l = 1,2. \tag{5.26}$$

For kernel weights defined according to (5.23) with kernels $K^{(l)}$ assume that the bandwidths satisfy

$$0 < h_n^{(l)} \to 0, \quad nh_n^{(l)d} \to \infty, \quad l = 1,2, \tag{5.27}$$

($K^{(l)}$ symmetric, $1_{S_{0,R}}(x) \geq K^{(l)}(x) \geq b1_{S_{0,r}}(x)$ $(0 < r \leq R < \infty,\ b > 0)$). For nearest neighbor weights defined according to (5.24), which refer only to $\widetilde{m}_n^{(LA)}(x)$ assume that

$$2 \leq k_n^{(2)} \leq n, \quad k_n^{(2)} \to \infty, \quad \frac{k_n^{(2)}}{n} \to 0 \tag{5.28}$$

Then for the estimate (5.21) under the above assumptions

$$\int \left|\widetilde{\sigma}_n^2(x)^{(LA)} - \sigma^2(x)\right| \mu(dx) \xrightarrow{P} 0$$

holds.
(Consistency of the local averaging estimator of the local variance under censoring)

Proof. We apply Lemma 3.2 with Y_i, $\overline{Y}_i$, m_n and m replaced by $\widetilde{Y_i^2} - m^2(X_i)$, $\widetilde{Y_{n,i}^2} - \widetilde{m}_n^2(X_i)^{(LA)}$, $\widetilde{\sigma}_n^{2\ (LA)}$ and σ^2, respectively. In order to do that, because of (5.6) it is enough to show

$$\frac{1}{n}\sum_{i=1}^{n}|(\widetilde{Y_{n,i}^2} - \widetilde{m}_n^2(X_i)^{(LA)}) - (\widetilde{Y_i^2} - m^2(X_i))| \xrightarrow{P} 0,$$

It suffices to show

$$\frac{1}{n}\sum_{i=1}^{n}|\widetilde{Y_{n,i}^2} - \widetilde{Y_i^2}| \xrightarrow{P} 0$$

and

$$\frac{1}{n}\sum_{i=1}^{n}|\widetilde{m}_n^2(X_i)^{(LA)} - m^2(X_i)| \xrightarrow{P} 0.$$

But

$$\frac{1}{n}\sum_{i=1}^{n}|\widetilde{Y_{n,i}^2} - \widetilde{Y_i^2}| = \frac{1}{n}\sum_{i=1}^{n}\delta_i T_i^2 \left|\frac{1}{G(T_i)} - \frac{1}{G_n(T_i)}\right| \to 0, \tag{5.29}$$

almost surely, by **(A2)** and Lemma 26.1 in [11] and thus in probability. For $\widetilde{m}_n^{(LA)}$ we notice that by Lemma 24.7 (Hardy-Littlewood) and its extension (24.10) in [11], pp. 503, 504 and Lemma 24.11 in [11] (for the empirical measure with respect to $(X_1, \ldots, X_n)$) and the function $x_i \to \widetilde{y}_{n,i} - \widetilde{y}_i$ ($i = 1, \ldots, n$) for the realizations $(x_i, \widetilde{y}_{n,i}, \widetilde{y}_i)$ of $(X_i, \widetilde{Y}_{n,i}, \widetilde{Y}_i)$, without sup and therefore without the assumption that the partitioning sequence is nested)

$$\frac{1}{n}\sum_{i=1}^{n}\left|\frac{\frac{1}{n}\sum_{j=1}^{n}(\widetilde{Y}_{n,j} - \widetilde{Y}_j)K\left(\frac{X_i - X_j}{h_n}\right)}{\frac{1}{n}K\left(\frac{X_i - X_j}{h_n}\right)}\right|^2 \le c^*\frac{1}{n}\sum_{i=1}^{n}|\widetilde{Y}_{n,i} - \widetilde{Y}_i|^2$$

and

$$\frac{1}{n}\sum_{i=1}^{n}\left|\frac{\frac{1}{n}\sum_{j=1}^{n}(\widetilde{Y}_{n,j} - \widetilde{Y}_j)1_{A_n(X_j)}(X_i)}{\frac{1}{n}1_{A_n(X_j)}(X_i)}\right|^2 \le c^*\frac{1}{n}\sum_{i=1}^{n}|\widetilde{Y}_{n,i} - \widetilde{Y}_i|^2$$

respectively, for some finite constant c^*, thus, setting

$$\widehat{m}_n(x) = \frac{\sum_{j=1}^{n}\widetilde{Y}_j K\left(\frac{x - X_j}{h_n}\right)}{\sum_{j=1}^{n}K\left(\frac{x - X_j}{h_n}\right)}$$

and

$$\widehat{m}_n(x) = \frac{\sum_{j=1}^n \widetilde{Y}_j 1_{A_n(x)}(X_j)}{\sum_{j=1}^n 1_{A_n(x)}(X_j)},$$

respectively,

$$\frac{1}{n}\sum_{i=1}^n |\widetilde{m}_n^{(LA)}(X_i) - \widehat{m}_n(X_i)|^2 \le c^* \frac{1}{n}\sum_{i=1}^n |\widetilde{Y}_{n,i} - \widetilde{Y}_i|^2 \to 0, \tag{5.30}$$

by **(A2)** and [22], Lemma 2, almost surely and thus in probability (by (5.29)).
Further

$$\begin{aligned} &\boldsymbol{E}\left\{\frac{1}{n}\sum_{i=1}^n |\widehat{m}_n(X_i) - m(X_i)|\right\} \\ &= \boldsymbol{E}\left\{|\widehat{m}_n(X_1) - m(X_1)|\right\} \\ &\quad \text{(by symmetry)} \\ &\quad \to 0, \end{aligned} \tag{5.31}$$

by Lemma 3.3, having replaced there Y_i by $\widetilde{Y}_i$.
Because of **(A2)** and (5.4), the random variables $\widetilde{Y}_i$ and $\hat{m}_n(X_i)$ are uniformly bounded. This together with (5.31) yields

$$\frac{1}{n}\sum_{i=1}^n |\hat{m}_n(X_i) - m(X_i)|^2 \xrightarrow{P} 0,$$

therefore, by (5.30)

$$\frac{1}{n}\sum_{i=1}^n |\widetilde{m}_n^{(LA)}(X_i) - m(X_i)|^2 \xrightarrow{P} 0. \tag{5.32}$$

Using the inequality

$$2ab \le \alpha a^2 + \frac{1}{\alpha} b^2$$

for arbitrary $a, b \in \mathbb{R}$, $\alpha > 0$, we obtain

$$\begin{aligned} &\frac{1}{n}\sum_{i=1}^n |\widetilde{m}_n^2(X_i)^{(LA)} - m^2(X_i)| \xrightarrow{P} 0 \\ &\le U^* \frac{1}{n}\sum_{i=1}^n |\widetilde{m}_n(X_i)^{(LA)} - m(X_i)| \xrightarrow{P} 0, \end{aligned}$$

for some random variable U^*.
Thus the assertion of the Theorem 5.3 is proven. □

The next theorem gives the rate of convergence for the estimator in (5.21).

Theorem 5.4. *Let the assumptions **(A1)-(A3)** hold. Let the estimate $\sigma_n^{2\,(LA)}$ be given by (5.21) with weights $W_{n,i}$ as in (5.23) and naive kernel $1_{S_{0,1}}$ with bandwidth $h_n \sim n^{-\frac{1}{d+2}}$. Moreover let m and σ^2 be Lipschitz continuous, that is*

$$|m(x) - m(z)| \leq \Gamma \|x - z\| \quad x,\ z \in \mathbb{R}^d$$

and

$$|\sigma^2(x) - \sigma^2(z)| \leq \Lambda \|x - z\| \quad x,\ z \in \mathbb{R}^d,$$

($\Lambda,\ \Gamma \in \mathbb{R}_+$, $\|\ \|$ denoting the Euclidean norm).
Then

$$\int |\widetilde{\sigma}_n^2(x)^{(LA)} - \sigma^2(x)|\mu(dx) = O_P\left(\left(\frac{\log n}{n}\right)^{\frac{1}{6}} + n^{-\frac{1}{d+2}}\right)$$

(Rate of convergence of the local averaging estimator of the local variance under censoring)

Proof. Introduce the following modification of the estimator (5.21)

$$\widehat{\widetilde{\sigma}}_n^2(x)^{(LA)} := \frac{\sum_{i=1}^n \widetilde{Z}_{n,i} K\left(\frac{x-X_i}{h_n}\right)}{n\mu(x + h_n S)}$$

Now, notice that

$$\begin{aligned}
&\int |\widetilde{\sigma}_n^2(x)^{(LA)} - \sigma^2(x)|\mu(dx) \\
&\leq \int |\widetilde{\sigma}_n^2(x)^{(LA)} - \widehat{\widetilde{\sigma}}_n^2(x)^{(LA)}|\mu(dx) + \int |\widehat{\widetilde{\sigma}}_n^2(x)^{(LA)} - \sigma^2(x)|\mu(dx) \\
&= A_n + B_n.
\end{aligned} \tag{5.33}$$

But, remembering (5.6), (compare [11], pp. 484, 485)

$$\begin{aligned}
A_n &= \int |\widetilde{\sigma}_n^2(x)^{(LA)} - \widehat{\widetilde{\sigma}}_n^2(x)^{(LA)}|\mu(dx) \\
&= U \int \left| \frac{\sum_{i=1}^n \widetilde{Z}_{n,i} K\left(\frac{x-X_i}{h_n}\right)}{\sum_{l=1}^n K\left(\frac{x-X_l}{h_n}\right)} - \frac{\sum_{i=1}^n \widetilde{Z}_{n,i} K\left(\frac{x-X_i}{h_n}\right)}{n\mu(x + h_n S)} \right| \mu(dx)
\end{aligned}$$

$$\leq U \int \sum_{i=1}^{n} K\left(\frac{x-X_i}{h_n}\right)$$

$$\left| \frac{1}{\sum_{l=1}^{n} K\left(\frac{x-X_l}{h_n}\right)} - \frac{1}{n\mu(x+h_nS)} \right| \mu(dx)$$

$$\leq U \int \left| 1 - \frac{\sum_{i=1}^{n} K\left(\frac{x-X_i}{h_n}\right)}{n\mu(x+h_nS)} \right| \mu(dx).$$

Now, from

$$\int \left| 1 - \frac{\sum_{i=1}^{n} K\left(\frac{x-X_i}{h_n}\right)}{n\mu(x+h_nS)} \right|^2 \mu(dx)$$

$$= O_P\left(h_n^2 + \frac{1}{nh_n^d}\right)$$

(having used Theorem 5.2 and its proof in [11] and the structure of h_n,)

we get

$$\int \left| 1 - \frac{\sum_{i=1}^{n} K\left(\frac{x-X_i}{h_n}\right)}{n\mu(x+h_nS)} \right| \mu(dx) = O_P\left(n^{-\frac{1}{d+2}}\right)$$

Therefore

$$\int |\widetilde{\sigma}_n^2(x)^{(LA)} - \widehat{\widetilde{\sigma}}_n^2(x^{(LA)})|\mu(dx) = O_P\left(n^{-\frac{1}{d+2}}\right).$$

For the term B_n in (5.33), we show again

$$\int |\widehat{\widetilde{\sigma}}_n^2(x^{(LA)}) - \sigma^2(x)|\mu(dx) = O_P\left(n^{-\frac{1}{d+2}}\right).$$

In fact

$$\int |\widehat{\widetilde{\sigma}}_n^2(x)^{(LA)} - \sigma^2(x)|\mu(dx)$$

$$\leq \int \left| \frac{\sum_{i=1}^{n} \left(\widetilde{Y_{n,i}^2} - \widetilde{m}_n^2(X_i)\right) K\left(\frac{x-X_i}{h_n}\right)}{n\mu(x+h_nS)} - \sigma^2(x) \right| \mu(dx)$$

$$
\begin{aligned}
&\leq \int \left| \frac{\sum_{i=1}^n \left(\widetilde{Y^2_{n,i}} - \widetilde{m}_n^2(X_i)\right) K\left(\frac{x-X_i}{h_n}\right)}{n\mu(x+h_nS)} \right. \\
&\qquad \left. - \frac{\sum_{i=1}^n \left(\widetilde{Y_i^2} - m^2(X_i)\right) K\left(\frac{x-X_i}{h_n}\right)}{n\mu(x+h_nS)} \right| \mu(dx) \\
&\quad + \int \left| \frac{\sum_{i=1}^n \left(\widetilde{Y_i^2} - m^2(X_i)\right) K\left(\frac{x-X_i}{h_n}\right)}{n\mu(x+h_nS)} - \sigma^2(x) \right| \mu(dx) \\
&= \int \left| \frac{\sum_{i=1}^n \left(\widetilde{Y^2_{n,i}} - \widetilde{m}_n^2(X_i) - \widetilde{Y_i^2} + m^2(X_i)\right) K\left(\frac{x-X_i}{h_n}\right)}{n\mu(x+h_nS)} \right| \mu(dx) \\
&\quad + \int \left| \frac{\sum_{i=1}^n \left(\widetilde{Y_i^2} - m^2(X_i)\right) K\left(\frac{x-X_i}{h_n}\right)}{n\mu(x+h_nS)} - \sigma^2(x) \right| \mu(dx) \\
&= C_n + D_n.
\end{aligned} \tag{5.34}
$$

Concerning C_n

$$
\begin{aligned}
&\int \left| \frac{\sum_{i=1}^n \left(\widetilde{Y^2_{n,i}} - \widetilde{m}_n^2(X_i) - \widetilde{Y_i^2} + m^2(X_i)\right) K\left(\frac{x-X_i}{h_n}\right)}{n\mu(x+h_nS)} \right| \mu(dx) \\
&\leq c\frac{1}{n}\sum_{i=1}^n \left| \left(\widetilde{Y^2_{n,i}} - \widetilde{m}_n^2(X_i)\right) - \left(\widetilde{Y_i^2} - m^2(X_i)\right) \right| \\
&\quad \text{(due to the Covering Lemma 23.6 in [11])} \\
&\leq c\frac{1}{n}\sum_{i=1}^n \left(|\widetilde{Y^2_{n,i}} - \widetilde{Y_i^2}| + |\widetilde{m}_n^2(X_i) - m^2(X_i)| \right) \\
&= c\left(E_n + F_n\right).
\end{aligned}
$$

But

$$
E_n = \frac{1}{n}\sum_{i=1}^n |\widetilde{Y^2_{n,i}} - \widetilde{Y_i^2}| = O_P\left(\left(\frac{\log n}{n} \right)^{\frac{1}{6}} \right),
$$

as in (5.19) under square root, and

$$
F_n = \frac{1}{n}\sum_{i=1}^n |\widetilde{m}_n^2(X_i) - m^2(X_i)|
$$

$$= \frac{1}{n}\sum_{i=1}^{n}|\widetilde{m}_n(X_i) + m(X_i)||\widetilde{m}_n(X_i) - m(X_i)|$$

$$\leq U^* \frac{1}{n}\sum_{i=1}^{n}|\widetilde{m}_n(X_i) - m(X_i)|$$

$$\leq \frac{U^*}{n}\sum_{i=1}^{n}\left|\frac{\sum_{j=1}^{n}\widetilde{Y}_{n,j}K\left(\frac{X_i-X_j}{h_n}\right)}{\sum_{j=1}^{n}K\left(\frac{X_i-X_j}{h_n}\right)} - \frac{\sum_{j=1}^{n}\widetilde{Y}_{j}K\left(\frac{X_i-X_j}{h_n}\right)}{\sum_{j=1}^{n}\widetilde{Y}_{n,j}K\left(\frac{X_i-X_j}{h_n}\right)}\right|$$

$$+\frac{U^*}{n}\sum_{i=1}^{n}\left|\frac{\sum_{j=1}^{n}\widetilde{Y}_{j}K\left(\frac{X_i-X_j}{h_n}\right)}{\sum_{j=1}^{n}\widetilde{Y}_{n,j}K\left(\frac{X_i-X_j}{h_n}\right)} - m(X_i)\right| = G_n + H_n.$$

Concerning G_n

$$\frac{1}{n}\sum_{i=1}^{n}\left|\frac{\sum_{j=1}^{n}\widetilde{Y}_{n,j}K\left(\frac{X_i-X_j}{h_n}\right)}{\sum_{j=1}^{n}K\left(\frac{X_i-X_j}{h_n}\right)} - \frac{\sum_{j=1}^{n}\widetilde{Y}_{j}K\left(\frac{X_i-X_j}{h_n}\right)}{\sum_{j=1}^{n}\widetilde{Y}_{n,j}K\left(\frac{X_i-X_j}{h_n}\right)}\right|$$

$$\leq \sqrt{\frac{1}{n}\sum_{i=1}^{n}\left|\frac{\sum_{j=1}^{n}\widetilde{Y}_{n,j}K\left(\frac{X_i-X_j}{h_n}\right)}{\sum_{j=1}^{n}K\left(\frac{X_i-X_j}{h_n}\right)} - \frac{\sum_{j=1}^{n}\widetilde{Y}_{j}K\left(\frac{X_i-X_j}{h_n}\right)}{\sum_{j=1}^{n}\widetilde{Y}_{n,j}K\left(\frac{X_i-X_j}{h_n}\right)}\right|^2}$$

(by the Cauchy-Schwarz inequality)

$$\leq \sqrt{c\frac{1}{n}\sum_{i=1}^{n}\left(\widetilde{Y}_{n,i} - \widetilde{Y}_i\right)^2} = \sqrt{O_P\left(\left(\frac{\log n}{n}\right)^{\frac{1}{3}}\right)}$$

$$= O_P\left(\left(\frac{\log n}{n}\right)^{\frac{1}{6}}\right) \tag{5.35}$$

Now, regarding H_n,

$$H_n \leq \frac{1}{n}\sum_{i=1}^{n}\left|\frac{\sum_{j=1}^{n}\widetilde{Y}_{j}K\left(\frac{X_i-X_j}{h_n}\right)}{\sum_{j=1}^{n}K\left(\frac{X_i-X_j}{h_n}\right)} - m(X_i)\right|$$

$$= \frac{1}{n}\sum_{i=1}^{n}\left|\widehat{m}_n(X_i) - m(X_i)\right|.$$

But

$$\boldsymbol{E}\left|\widehat{m}_n(X_1) - m(X_1)\right|^2 = O\left(n^{-\frac{2}{d+2}}\right),$$

because of

$$\int \frac{1}{n\mu(S_{x,h_n})}\mu(dx) = O\left(\frac{1}{nh_n^d}\right) = O\left(n^{-\frac{2}{d+2}}\right)$$

and using the argument of Lemma 3.3.
Therefore

$$H_n = O_P\left(n^{-\frac{1}{d+2}}\right).$$

This, together with (5.35) implies

$$C_n = O_P\left(\left(\frac{\log n}{n}\right)^{\frac{1}{6}} + n^{-\frac{1}{d+2}}\right)$$

To get the assertion it remains to establish a rate of convergence for the term D_n in (5.34)

$$\int \left| \frac{\sum_{i=1}^n \left(\widetilde{Y_i^2} - m^2(X_i)\right) K\left(\frac{x-X_i}{h_n}\right)}{n\mu(x+h_nS)} - \sigma^2(x) \right| \mu(dx).$$

But under the Lipschitz continuity of σ^2, it is equivalent to the problem of establishing a rate of convergence for

$$\int \left| \frac{\sum_{i=1}^n Y_i K\left(\frac{x-X_i}{h_n}\right)}{n\mu(x+h_nS)} - m(x) \right| \mu(dx),$$

having replaced there Y_i by $Y_i^2 - m^2(X_i)$ and m by σ^2.
Therefore, according to Theorem 5.2 in [11] and its proof it holds that $D_n = O_P\left(n^{-\frac{1}{d+2}}\right)$.
Thus the assertion is obtained. □

Remark 5.2. Let the estimate be given by (5.21) with weights as in (5.22) of cubic partition with side length $h_n = n^{-\frac{1}{d+2}}$. Then an analogous result as in Theorem 5.4 holds, noticing that

$$\int \frac{1}{n\mu(A_n(x))}\mu(dx) = O\left(\frac{1}{nh_n^d}\right) = O\left(n^{-\frac{1}{d+2}}\right).$$

5.4 Censored Partitioning Estimation via Nearest Neighbors

In this section, analogously to chapter 4, the aim is to discuss estimators of the local variance function with partitioning approach, based on the first and second nearest neighbors. Moreover here the crucial fact is the additional modeling of censoring. The treatment is more difficult as in uncensored case, therefore we need some helpful lemmas, that we will present in short. Before them, recall Section 5.1, the definitions of nearest neighbors, (4.4) and (4.5), and define, for $i = 1, \dots, n$

$$\delta_{N[i,k]} = 1_{\{Y_{N[i,k]} \le C_i\}}$$

and

$$T_{N[i,k]} = \min(Y_{N[i,k]}, C_i).$$

Finally, assume that ties occur with probability zero. (If ties occur, there is different possibility to break them, as explained in Section 4.1. The possibility to break them provide for simplicity the assumption that they occur with probability zero.)
Now, tree helpful lemmas.

Lemma 5.1. *With the above definitions and Definition 5.1, it holds*

$$\boldsymbol{E}\left\{ \frac{\delta_i T_i}{G(T_i)} \frac{\delta_{N[i,1]} T_{N[i,1]}}{G(T_{N[i,1]})} \middle| X_i \right\} = \boldsymbol{E}\left\{ Y_i Y_{N[i,1]} | X_i \right\}. \tag{5.36}$$

Proof. Consider that

$$\begin{aligned}
&\boldsymbol{E}\left\{ \frac{\delta_i T_i}{G(T_i)} \frac{\delta_{N[i,1]} T_{N[i,1]}}{G(T_{N[i,1]})} \middle| X_1, \dots, X_n \right\} \\
&= \boldsymbol{E}\left\{ \sum_{l \in \{1,\dots,n\} \setminus \{i\}} \frac{\delta_i T_i}{G(T_i)} \frac{\delta_l T_l}{G(T_l)} 1_{\{N[i,1]=l\}} \middle| X_1, \dots, X_n \right\} \\
&= \sum_{l \in \{1,\dots,n\} \setminus \{i\}} \boldsymbol{E}\left\{ \frac{\delta_i T_i}{G(T_i)} \frac{\delta_l T_l}{G(T_l)} 1_{\{N[i,1]=l\}} \middle| X_1, \dots, X_n \right\} \\
&= \sum_{l \in \{1,\dots,n\} \setminus \{i\}} \boldsymbol{E}\left\{ \frac{\delta_i T_i}{G(T_i)} \middle| X_i \right\} \boldsymbol{E}\left\{ \frac{\delta_l T_l}{G(T_l)} \middle| X_l \right\} 1_{\{N[i,1]=l\}}
\end{aligned}$$

(by the independence assumption)

$$= \boldsymbol{E}\left\{ \frac{\delta_i T_i}{G(T_i)} \middle| X_i \right\} \sum_{l \in \{1,\ldots,n\}\setminus\{i\}} \boldsymbol{E}\left\{ \frac{\delta_l T_l}{G(T_l)} \middle| X_l \right\} 1_{\{N[i,1]=l\}}$$

$$= \boldsymbol{E}\{Y_i | X_i\} \sum_{l \in \{1,\ldots,n\}\setminus\{i\}} \boldsymbol{E}\{Y_l | X_l\} 1_{\{N[i,1]=l\}},$$

the latter by (5.7).
Moreover

$$\begin{aligned}
&\boldsymbol{E}\left\{Y_i Y_{N[i,1]} | X_1, \ldots, X_n\right\} \\
&= \boldsymbol{E}\left\{ \sum_{l \in \{1,\ldots,n\}\setminus\{i\}} Y_i Y_l 1_{\{N[i,1]=l\}} \middle| X_1, \ldots, X_n \right\} \\
&= \sum_{l \in \{1,\ldots,n\}\setminus\{i\}} \boldsymbol{E}\left\{Y_i Y_l 1_{\{N[i,1]=l\}} | X_1, \ldots, X_n\right\} \\
&= \sum_{l \in \{1,\ldots,n\}\setminus\{i\}} \boldsymbol{E}\left\{Y_i Y_l | X_1, \ldots, X_n\right\} 1_{\{N[i,1]=l\}} \\
&= \sum_{l \in \{1,\ldots,n\}\setminus\{i\}} \boldsymbol{E}\left\{Y_i | X_i\right\} \boldsymbol{E}\left\{Y_l | X_l\right\} 1_{\{N[i,1]=l\}} \\
&\quad \text{(by independence)} \\
&= \boldsymbol{E}\left\{Y_i | X_i\right\} \sum_{l \in \{1,\ldots,n\}\setminus\{i\}} \boldsymbol{E}\left\{Y_l | X_l\right\} 1_{\{N[i,1]=l\}}.
\end{aligned}$$

These results imply (5.36). □

Analogously to the above lemma one has

Lemma 5.2. *It holds*

$$\boldsymbol{E}\left\{ \frac{\delta_i T_i}{G(T_i)} \frac{\delta_{N[i,2]} T_{N[i,2]}}{G(T_{N[i,2]})} \middle| X_i \right\} = \boldsymbol{E}\left\{Y_i Y_{N[i,2]} | X_i\right\}. \tag{5.37}$$

The proof is analogous to the proof of Lemma 5.1 and therefore omitted. A similar argument yields the following

Lemma 5.3. *It holds*

$$\boldsymbol{E}\left\{ \frac{\delta_{N[i,1]} T_{N[i,1]}}{G(T_{N[i,1]})} \frac{\delta_{N[i,2]} T_{N[i,2]}}{G(T_{N[i,2]})} \middle| X_i \right\} = \boldsymbol{E}\left\{Y_{N[i,1]} Y_{N[i,2]} | X_i\right\}. \tag{5.38}$$

Again, the proof is omitted.

Recall then the following known relation (see (5.8))

$$\boldsymbol{E}\left\{\frac{\delta_i T_i^2}{G(T_i)}\right\} = \boldsymbol{E}\{Y_i^2|X_i\}. \tag{5.39}$$

Set now

$$\begin{aligned} H_i &:= H_{n,i} \\ &:= \frac{\delta_i T_i^2}{G(T_i)} - \frac{\delta_i T_i}{G(T_i)} \frac{\delta_{N[i,1]} T_{N[i,1]}}{G(T_{N[i,1]})} - \frac{\delta_i T_i}{G(T_i)} \frac{\delta_{N[i,2]} T_{N[i,2]}}{G(T_{N[i,2]})} \\ &\quad + \frac{\delta_{N[i,1]} T_{N[i,1]}}{G(T_{N[i,1]})} \frac{\delta_{N[i,2]} T_{N[i,2]}}{G(T_{N[i,2]})} \end{aligned} \tag{5.40}$$

for $i \in \{1, \ldots, n\}$ and note

$$\begin{aligned} &\boldsymbol{E}\{H_i|X_i = x\} \\ &= \boldsymbol{E}\left\{Y_i^2 - Y_i Y_{N[i,1]} - Y_i Y_{N[i,2]} + Y_{N[i,1]} Y_{N[i,2]} | X_i = x\right\} \\ &\quad \text{(the latter by Lemmas 5.1, 5.2 and 5.3)} \\ &= \boldsymbol{E}\left\{(Y_i - Y_{N[i,1]})(Y_i - Y_{N[i,2]})|X_i = x\right\} = \boldsymbol{E}\left\{W_i|X_i = x\right\} \end{aligned} \tag{5.41}$$

with

$$W_i := (Y_i - m(X_i))^2 + (m(X_i) - m(X_{N[i,1]}))(m(X_i) - m(X_{N[i,2]})) \tag{5.42}$$

according to Liitiäinen at al. ([19], [18]).
Our proposal for an estimator of the local variance function under known survival function G is given by

$$\widehat{\sigma}_n^2(x) := \frac{\sum_{i=1}^n H_i 1_{A_n(x)}(X_i)}{\sum_{i=1}^n 1_{A_n(x)}(X_i)}. \tag{5.43}$$

The following theorem states consistency of this estimator.

Theorem 5.5. *Let Assumptions **(A1)-(A3)** hold.*
Let $\mathcal{P}_n = \{\boldsymbol{A}_{n,1}, \ldots, \boldsymbol{A}_{n,l_n}\}$ be a sequence of partitions on $\mathbb{R}^d$ such that for each sphere S centered at the origin

$$\lim_{n\to\infty} \max_{j\in\{A_{n,j}\cap S\neq\emptyset\}} \operatorname{diam} \boldsymbol{A}_{n,j} = 0, \tag{5.44}$$

and

$$\lim_{n\to\infty} \frac{\#\{j:\ A_{n,j}\cap S \neq \emptyset\}}{n} = 0. \tag{5.45}$$

Then

$$\int |\widehat{\sigma}_n^2 - \sigma^2(x)|\mu(dx) \xrightarrow{P} 0.$$

(Consistency of the partitioning estimator of the local variance under censoring and known survival function)

Before giving the proof of this theorem, introduce the following modification of the estimation (5.43)

$$\widehat{\widehat{\sigma}}_n^2(x) := \frac{\sum_{i=1}^n H_i 1_{A_n(x)}(X_i)}{n\mu(A_n(x))}, \tag{5.46}$$

and the following lemma.

Lemma 5.4. *Under the conditions of Theorem 5.5,* $\widehat{\widehat{\sigma}}_n^2(x)$ *is consistent, i.e.,*

$$\int |\widehat{\widehat{\sigma}}_n^2(x) - \sigma^2(x)|\mu(dx) \xrightarrow{P} 0.$$

Proof. Choose a sphere S centered at 0 which contains the support of μ. Set $J_n := \{j : A_{n,j} \cap S \neq \emptyset\}$ and $l_n := \# J_n$. The variance of the estimator can be bounded by

$$\boldsymbol{Var}\left\{\widehat{\widehat{\sigma}}_n^2(x)\right\} \leq 72\frac{L^4}{G(L)^4}\frac{1}{n\mu(A_n(x))}.$$

he notation $N'[i,1]$ $N'[i,2]$ underling that we are searching the neighbors from the set $\{X_1', X_2, \ldots, X_n\} \setminus \{X_i\}$ to achieve $T'_{N[i,1]}$ and $T'_{N[i,2]}$, respectively.
It holds

$$\begin{aligned}
&\boldsymbol{Var}\left\{\widehat{\widehat{\sigma}}_n^2(x)\right\} \\
&\leq \frac{4}{n^2\mu(A_n(x))^2}\left[\boldsymbol{Var}\left\{\sum_{i=1}^n \frac{\delta_i T_i^2 1_{A_n(x)}(X_i)}{G(T_i)}\right\}\right. \\
&+\boldsymbol{Var}\left\{\sum_{i=1}^n \frac{\delta_i T_i}{G(T_i)}\frac{\delta_{N[i,1]}T_{N[i,1]}1_{A_n(x)}(X_i)}{G(T_{N[i,1]})}\right\} \\
&+\boldsymbol{Var}\left\{\sum_{i=1}^n \frac{\delta_1 T_1}{G(T_1)}\frac{\delta_{N[i,2]}T_{N[i,2]}1_{A_n(x)}(X_i)}{G(T_{N[i,2]})}\right\} \\
&\left.+\boldsymbol{Var}\left\{\sum_{i=1}^n \frac{\delta_{N[i,1]}T_{N[i,1]}1_{A_n(x)}(X_i)}{G(T_{N[i,1]})}\frac{\delta_{N[i,2]}T_{N[i,2]}1_{A_n(x)}(X_i)}{G(T_{N[i,2]})}\right\}\right].
\end{aligned}$$

Each of the four variances in the right-hand side is bounded by

$$18\frac{L^4}{G(L)^4}n\mu(A_n(x)).$$

We show this only for the fourth variance because the other three variances can be treated in the same way. We apply the Efron-Stein inequality, following the argument in the proof of (4.18).
Let $n \geq 2$ be fixed. Replacement of (X_j, Y_j, C_j) by (X_j', Y_j', C_j') for fixed $j \in \{1, \ldots, n\}$ (where $(X_1, Y_1, C_1), \ldots, (X_n, Y_n, C_n)$, $(X_1', Y_1', C_1'), \ldots, (X_n', Y_n', C_n')$ are independent and identically distributed) leads, for fixed x, from

$$U_n := \sum_{i=1}^{n} \frac{\delta_{N[i,1]} T_{N[i,1]} 1_{A_n(x)}(X_i)}{G(T_{N[i,1]})} \cdot \frac{\delta_{N[i,2]} T_{N[i,2]} 1_{A_n(x)}(X_i)}{G(T_{N[i,2]})},$$

$N[j,1]$ and $N[j,2]$ to $U_{n,j}$, $N'[j,1]$ and $N'[j,2]$, respectively.
With $T_j' := \min\left\{Y_j', C_j'\right\}$, $\delta_j' = 1_{\{Y_j' \leq C_j'\}}$ we obtain

$$|U_n - U_{n,j}| \leq A_{n,j} + B_{n,j} + C_{n,j} + D_{n,j} + E_{n,j} + F_{n,j}$$

where with $Z_i = \frac{\delta_i T_i}{G(T_i)}$, $Z_j' = \frac{\delta_j' T_j'}{G(T_j')}$ and (5.4)

$$\begin{aligned}
A_{n,j} &= \sum_{\substack{l,\ q \in \{1,\ldots,n\}\setminus\{j\} \\ l \neq q}} Z_l Z_q 1_{A_n(x)}(X_i) 1_{\{N[j,1]=l\}} 1_{\{N[j,2]=q\}} \\
&\leq \frac{L^2}{G(L)^2} 1_{A_n(x)}(X_i), \\
B_{n,j} &= \sum_{\substack{l,\ q \in \{1,\ldots,n\}\setminus\{j\} \\ l \neq q}} Z_l Z_q 1_{A_n(x)}(X_i') 1_{\{N'[j,1]=l\}} 1_{\{N'[j,2]=q\}} \\
&\leq \frac{L^2}{G(L)^2} 1_{A_n(x)}(X_i'), \\
C_{n,j} &= \sum_{\substack{i,\ q \in \{1,\ldots,n\}\setminus\{j\} \\ i \neq q}} Z_j Z_q 1_{A_n(x)}(X_i) 1_{\{N[i,1]=j\}} 1_{\{N[i,2]=q\}} \\
&\leq \frac{L^2}{G(L)^2} \sum_{i \in \{1,\ldots,n\}\setminus\{j\}} 1_{A_n(x)}(X_i) 1_{\{N[i,1]=j\}},
\end{aligned}$$

$$D_{n,j} = \sum_{\substack{i,\ q\in\{1,\ldots,n\}\setminus\{j\} \\ i\neq q}} Z_j' Z_q 1_{A_n(x)}(X_i) 1_{\{N'[i,1]=j\}} 1_{\{N'[i,2]=q\}},$$
$$\leq \frac{L^2}{G(L)^2} \sum_{i\in\{1,\ldots,n\}\setminus\{j\}} 1_{A_n(x)}(X_i) 1_{\{N'[i,1]=j\}},$$
$$E_{n,j} = \sum_{\substack{i,\ l\in\{1,\ldots,n\}\setminus\{j\} \\ i\neq l}} Z_l Z_j 1_{A_n(x)}(X_i) 1_{\{N[i,1]=l\}} 1_{\{N[i,2]=j\}},$$
$$\leq \frac{L^2}{G(L)^2} \sum_{i\in\{1,\ldots,n\}\setminus\{j\}} 1_{A_n(x)}(X_i) 1_{\{N[i,2]=j\}},$$
$$F_{n,j} = \sum_{\substack{i,\ l\in\{1,\ldots,n\}\setminus\{j\} \\ i\neq l}} Z_l Z_j' 1_{A_n(x)}(X_i) 1_{\{N'[i,1]=l\}} 1_{\{N'[i,2]=j\}}$$
$$\leq \frac{L^2}{G(L)^2} \sum_{i\in\{1,\ldots,n\}\setminus\{j\}} 1_{A_n(x)}(X_i) 1_{\{N'[i,2]=j\}}.$$

Now,

$$A_{n,j}^2 \leq \frac{L^4}{G(L)^4} 1_{A_n(x)}(X_i),$$
$$B_{n,j}^2 \leq \frac{L^4}{G(L)^4} 1_{A_n(x)}(X_i'),$$
$$C_{n,j}^2 \leq \frac{L^4}{G(L)^4} \sum_{i\in\{1,\ldots,n\}\setminus\{j\}} 1_{A_n(x)}(X_i) 1_{\{N[i,1]=j\}},$$

(by the Cauchy-Schwarz inequality)

$$D_{n,j}^2 \leq \frac{L^4}{G(L)^4} \sum_{i\in\{1,\ldots,n\}\setminus\{j\}} 1_{A_n(x)}(X_i) 1_{\{N'[i,1]=j\}},$$
$$E_{n,j}^2 \leq \frac{L^4}{G(L)^4} \sum_{i\in\{1,\ldots,n\}\setminus\{j\}} 1_{A_n(x)}(X_i) 1_{\{N[i,2]=j\}},$$
$$F_{n,j}^2 \leq \frac{L^4}{G(L)^4} \sum_{i\in\{1,\ldots,n\}\setminus\{j\}} 1_{A_n(x)}(X_i) 1_{\{N'[i,2]=j\}}.$$

Considering now the terms $\sum_{j=1}^n \mathbf{E}\left\{A_{n,j}^2\right\}$ and $\sum_{j=1}^n \mathbf{E}\left\{B_{n,j}^2\right\}$, we have for them an upper bound

$$\frac{L^4}{G(L)^4} n\mu(A_n(x)),$$

respectively. Analogously, considering the terms $\sum_{j=1}^n \boldsymbol{E}\left\{C_{n,j}^2\right\}$, $\sum_{j=1}^n \boldsymbol{E}\left\{D_{n,j}^2\right\}$, $\sum_{j=1}^n \boldsymbol{E}\left\{E_{n,j}^2\right\}$ and $\sum_{j=1}^n \boldsymbol{E}\left\{F_{n,j}^2\right\}$, by changing the order of summation, for each of these terms we have an upper bound

$$\frac{L^4}{G(L)^4}\boldsymbol{E}\left\{\sum_{i\in\{1,\ldots,n\}} 1_{A_n(x)}(X_i)\right\} \le \frac{L^4}{G(L)^4} n\mu(A_n(x)).$$

Thus

$$\begin{aligned}
&\boldsymbol{E}\left\{\sum_{j=1}^n |U_n - U_{n,j}|^2\right\}\\
&\le 6\boldsymbol{E}\left(\sum_{j=1}^n \boldsymbol{E}\left\{A_{n,j}^2\right\} + \sum_{j=1}^n \boldsymbol{E}\left\{B_{n,j}^2\right\} + \sum_{j=1}^n \boldsymbol{E}\left\{C_{n,j}^2\right\}\right.\\
&\quad \left. + \sum_{j=1}^n \boldsymbol{E}\left\{D_{n,j}^2\right\} + \sum_{j=1}^n \boldsymbol{E}\left\{E_{n,j}^2\right\} + \sum_{j=1}^n \boldsymbol{E}\left\{F_{n,j}^2\right\}\right)\\
&\le 6\cdot 6\frac{L^4}{G(L)^4} n\mu(A_n(x)),
\end{aligned}$$

which, by the Efron-Stein inequality, yields the above bound of the variance. Then, we have

$$\boldsymbol{Var}\left\{\widehat{\widehat{\sigma}}_n^2(x)\right\} \le \frac{72\cdot L^4}{G(L)^4}\frac{1}{n\mu(A_n(x))}.$$

By the well known relation for the mean squared error we get

$$\boldsymbol{E}\left\{|\widehat{\widehat{\sigma}}_n^2(x) - \boldsymbol{E}\widehat{\widehat{\sigma}}_n^2(x)|\right\} \le \sqrt{\boldsymbol{Var}(\widehat{\widehat{\sigma}}_n^2(x))} \le 6\sqrt{2}\frac{L^2}{G(L)^2}\frac{1}{\sqrt{n\mu(A_n(x))}}. \tag{5.47}$$

By the triangle inequality

$$\boldsymbol{E}|\widehat{\widehat{\sigma}}_n^2(x) - \sigma^2(x)| \le \boldsymbol{E}|\widehat{\widehat{\sigma}}_n^2(x) - \boldsymbol{E}\widehat{\widehat{\sigma}}_n^2(x)| + |\boldsymbol{E}\widehat{\widehat{\sigma}}_n^2(x) - \sigma^2(x)|, \tag{5.48}$$

and, with $l_n = \#\{j : A_{n,j}\cap S \ne \emptyset\}$ we note, with some constant c

$$\int_S \boldsymbol{E}|\widehat{\widehat{\sigma}}_n^2(x) - \boldsymbol{E}\widehat{\widehat{\sigma}}_n^2(x)|\mu(dx) \le c\frac{1}{\sqrt{n}}\int_S \frac{1}{\sqrt{\mu(A_n(x))}}\mu(dx)$$

$$\le c\frac{1}{\sqrt{n}}\sqrt{\int_S \frac{1}{\mu(A_n(x))}\mu(dx)} = O\left(\sqrt{\frac{l_n}{n}}\right) = o\left(\left(n_n^{-d}n^{-1}\right)^{\frac{1}{2}}\right). \tag{5.49}$$

Further

$$\begin{aligned}
\boldsymbol{E}\widehat{\widehat{\sigma}}_n^2(x) &= \frac{\boldsymbol{E}\left(H_1 1_{A_n(x)}(X_1)\right)}{\mu(A_n(x))} \\
&\text{(by symmetry)} \\
&= \frac{\boldsymbol{E}\left(\boldsymbol{E}\left(H_1 1_{A_n(x)}(X_1)\right)|X_1\right)}{\mu(A_n(x))} \\
&= \frac{\boldsymbol{E}\left(\boldsymbol{E}\left(H_1|X_1\right) 1_{A_n(x)}(X_1)\right)}{\mu(A_n(x))} \\
&= \frac{\boldsymbol{E}\left(\boldsymbol{E}\left(W_1|X_1\right) 1_{A_n(x)}(X_1)\right)}{\mu(A_n(x))} \\
&\text{(by 5.41)} \\
&= \int \frac{\boldsymbol{E}\{W_1|X_1 = z\}1_{A_n(x)}(z)}{\mu(A_n(x))}\mu(dz) \\
&= \boldsymbol{E}\sigma_n^{2*}(x)
\end{aligned} \tag{5.50}$$

(see the proof of Lemma 4.1 with $\sigma_n^{2*}(x)$ defined by (4.23)). Then, according to the proof of Lemma 4.1, we have

$$\begin{aligned}
&\int_S \left|\sigma^2(x) - \boldsymbol{E}\widehat{\widehat{\sigma}}_n^2(x)\right| \mu(dx) = K_n \\
&:= \int_S \left|\sigma^2(x) - \boldsymbol{E}\widehat{\widehat{\sigma}}_n^{2*}(x)\right| \mu(dx) \to 0
\end{aligned} \tag{5.51}$$

(5.48), together with (5.49) and (5.51) yield the assertion. □

Proof of Theorem 5.5. We begin by the following extension

$$\begin{aligned}
&\int |\widehat{\sigma}_n^2(x) - \sigma^2(x)|\mu(dx) \\
&\leq \int |\widehat{\sigma}_n^2(x) - \widehat{\widehat{\sigma}}_n^2(x)|\mu(dx) + \int |\widehat{\widehat{\sigma}}_n^2(x) - \sigma^2(x)|\mu(dx) \\
&\leq L_n + D_n.
\end{aligned}$$

It holds $D_n \xrightarrow{P} 0$ because of Lemma 5.4.
Now, concerning L_n, arguing as in Györfi et al. [11], p. 465, compare also the end of the proof of Theorem 4.3

$$\begin{aligned}
&\int |\widehat{\sigma}_n^2(x) - \widehat{\widehat{\sigma}}_n^2(x)|\mu(dx) \\
&\leq \int \left| \frac{\sum_{i=1}^n H_i 1_{A_n(x)}(X_i)}{\sum_{i=1}^n 1_{A_n(x)}(X_i)} - \frac{\sum_{i=1}^n H_i 1_{A_n(x)}(X_i)}{n\mu(A_n(x))} \right| \mu(dx) \\
&\leq const \int \sum_{i=1}^n 1_{A_n(x)}(X_i) \left| \frac{1}{n\mu(A_n(x))} - \frac{1}{\sum_{i=1}^n 1_{A_n(x)}(X_i)} \right| \mu(dx) \\
&\quad \text{(for some finite constant, because of } \mathbf{(A2)} \text{ and (5.4))} \\
&\leq const \int \left| \sum_{i=1}^n \frac{1_{A_n(x)}(X_i)}{n\mu(A_n(x))} - 1 \right| \mu(dx) \to 0
\end{aligned}$$

because of (5.44) and (5.45), which proves the theorem. □

As already treated in the previous section the survival function G is typically unknown and has to be estimated, by the Kaplan-Meier estimator. We introduce now the final result, in order to show consistency of the partitioning estimator of the local variance based on the first and the second neighbor under censoring and unknown survival function.
Let

$$\widetilde{\sigma}_n^2(x)^{(NN)} := \frac{\sum_{i=1}^n H_{i,G_n} 1_{A_n(x)}(X_i)}{\sum_{i=1}^n 1_{A_n(x)}(X_i)} \tag{5.52}$$

where

$$\begin{aligned}
H_{i,G_n} &:= H_{n,i,G_n} \\
&= \frac{\delta_i T_i^2}{G_n(T_i)} - \frac{\delta_i T_i}{G_n(T_i)} \frac{\delta_{N[i,1]} T_{N[i,1]}}{G_n(T_{N[i,1]})} - \frac{\delta_i T_i}{G_n(T_i)} \frac{\delta_{N[i,2]} T_{N[i,2]}}{G_n(T_{N[i,2]})} \\
&\quad + \frac{\delta_{N[i,1]} T_{N[i,1]}}{G_n(T_{N[i,1]})} \frac{\delta_{N[i,2]} T_{N[i,2]}}{G_n(T_{N[i,2]})}
\end{aligned}$$

Then for this estimator a consistency result holds; we prove this as follows.

Theorem 5.6. *Under the assumptions of Theorem 5.5,*

$$\int |\widetilde{\sigma}_n^2(x)^{(NN)} - \sigma^2(x)|\mu(dx) \xrightarrow{P} 0.$$

(Consistency of the partitioning estimator of the local variance under censoring)

Proof. Introduce the following modification of the estimator (5.52)

$$\widehat{\widetilde{\sigma}}_n^2(x)^{(NN)} := \frac{\sum_{i=1}^n H_{i,G_n} 1_{A_n(x)}(X_i)}{n\mu(A_n(x))}. \tag{5.53}$$

We note

$$\begin{aligned}
&\int |\widetilde{\sigma}_n^2(x)^{(NN)} - \sigma^2(x)|\mu(dx) \\
&\leq \int |\widetilde{\sigma}_n^2(x)^{(NN)} - \widehat{\widetilde{\sigma}}_n^2(x)^{(NN)}|\mu(dx) \\
&\quad + \int |\widehat{\widetilde{\sigma}}_n^2(x)^{(NN)} - \widehat{\widetilde{\sigma}}_n^2(x)|\mu(dx) \\
&\quad + \int |\widehat{\widetilde{\sigma}}_n^2(x) - \sigma^2(x)|\mu(dx) \\
&= A_n + B_n + C_n,
\end{aligned}$$

with $\widehat{\widetilde{\sigma}}_n^2$ defined by (5.46). But $C_n \xrightarrow{P} 0$ *a.s.* by Lemma 5.4.
Now, concerning A_n

$$\begin{aligned}
&\int |\widetilde{\sigma}_n^2(x)^{(NN)} - \widehat{\widetilde{\sigma}}_n^2(x)^{(NN)}|\mu(dx) \\
&\leq \frac{\sum_{i=1}^n H_{i,G_n} 1_{A_n(x)}(X_i)}{\sum_{i=1}^n 1_{A_n(x)}(X_i)} - \frac{\sum_{i=1}^n H_{i,G_n} 1_{A_n(x)}(X_i)}{n\mu(A_n(x))} \\
&\leq U^* \int \left| \frac{\sum_{i=1}^n 1_{A_n(x)(X_i)}}{n\mu(A_n(x))} \right| \mu(dx) \to 0, \quad a.s.
\end{aligned}$$

for some random variable $U^* < \infty$ (see [11], p. 465, by (5.6) and the boundedness of C).
Finally, concerning B_n

$$\int |\widehat{\widetilde{\sigma}}_n^2(x)^{(NN)} - \widehat{\widetilde{\sigma}}_n^2(x)|\mu(dx)$$

$$= \int \left| \frac{\sum_{i=1}^{n} H_{i,G_n} 1_{A_n(x)}(X_i)}{n\mu(A_n(x))} - \frac{\sum_{i=1}^{n} H_i 1_{A_n(x)}(X_i)}{n\mu(A_n(x))} \right| \mu(dx)$$

$$= \int \left| \frac{1}{n} \frac{\sum_{i=1}^{n} [H_{i,G_n} - H_i] 1_{A_n(x)}(X_i)}{\mu(A_n(x))} \right| \mu(dx)$$

$$\leq \frac{1}{n} \sum_{i=1}^{n} |H_{i,G_n} - H_i| \underbrace{\int \frac{1_{A_n(x)}(X_i)}{\mu(A_n(x))} \mu(dx)}_{\leq 1 \text{ because of } \mu(A_n(x)) = \mu(A_n(X_i)) \text{ for } X_i \in A_n(x)}. \tag{5.54}$$

But

$$\frac{1}{n} \sum_{i=1}^{n} |H_{i,G_n} - H_i|$$
$$\leq \frac{1}{n} \sum_{i=1}^{n} \left\{ \left| \frac{\delta_i T_i^2}{G_n(T_i)} - \frac{\delta_i T_i^2}{G(T_i)} \right| \right.$$
$$+ \left| \frac{\delta_i T_i}{G_n(T_i)} \frac{\delta_{N[i,1]} T_{N[i,1]}}{G_n(T_{N[i,1]})} - \frac{\delta_i T_i}{G(T_i)} \frac{\delta_{N[i,1]} T_{N[i,1]}}{G(T_{N[i,1]})} \right|$$
$$+ \left| \frac{\delta_i T_i}{G_n(T_i)} \frac{\delta_{N[i,2]} T_{N[i,2]}}{G_n(T_{N[i,2]})} - \frac{\delta_i T_i}{G(T_i)} \frac{\delta_{N[i,2]} T_{N[i,2]}}{G(T_{N[i,2]})} \right|$$
$$\left. + \left| \frac{\delta_{N[i,1]} T_{N[i,1]}}{G_n(T_{N[i,1]})} \frac{\delta_{N[i,2]} T_{N[i,2]}}{G_n(T_{N[i,2]})} - \frac{\delta_{N[i,1]} T_{N[i,1]}}{G(T_{N[i,1]})} \frac{\delta_{N[i,2]} T_{N[i,2]}}{G(T_{N[i,2]})} \right| \right\}$$
$$=: \frac{1}{n} \sum_{i=1}^{n} (P_{n,i} + O_{n,i} + I_{n,i} + U_{n,i}). \tag{5.55}$$

Now, concerning $P_{n,i}$,

$$\frac{1}{n} \sum_{i=1}^{n} P_{n,i} \leq \frac{L}{n} \sum_{i=1}^{n} \delta_i T_i \left| \frac{1}{G_n(T_i)} - \frac{1}{G(T_i)} \right| \to 0 \quad a.s.$$

due to Lemma 26.1 in [11].
Finally, concerning $U_{n,i}$, (and similarly, for $O_{n,i}$ and $I_{n,i}$) we recall (5.5) and notice that, for the ordered sequence of the variables of the first (and second) neighbors we get, with obvious meaning of the notation,

$$T_{N[(1),1]} \leq T_{N[(2),1]} \leq \cdots \leq T_{N[(n),1]} \leq T_K \leq L \quad a.s.,$$

$$T_{N[(1),2]} \leq T_{N[(2),2]} \leq \cdots \leq T_{N[(n),2]} \leq T_K \leq L \quad a.s.$$

and, with same positive random variable U^*

$$1 \geq G_n(T_{N[(1),1]}) \geq G_n(T_{N[(2),1]}) \geq \cdots \geq G_n(T_{N[(n),1]})$$
$$\geq G_n(T_K) \geq G_n(L) \geq U^* > 0 \quad a.s.,$$

$$1 \geq G_n(T_{N[(1),2]}) \geq G_n(T_{N[(2),2]}) \geq \cdots \geq G_n(T_{N[(n),2]})$$
$$\geq G_n(T_K) \geq G_n(L) \geq U^* > 0 \quad a.s., \tag{5.56}$$

respectively, because of $G_n(L) \to G(L) > 0$ *a.s.* by [11], Theorem 26.1. By this

$$H_{i,G_n} < U^{**} < \infty \quad (i = 1, \ldots, n,\ n \in \mathbb{N}) \quad a.s. \tag{5.57}$$

with some positive random variable U^{**}.
Now,

$$\begin{aligned}
&\left| \frac{1}{n} \sum_{i=1}^{n} U_{n,i} \right| \\
&= \left| \frac{1}{n} \sum_{i=1}^{n} \left[\frac{\delta_{N[i,1]} T_{N[i,1]}}{G_n(T_{N[i,1]})} \frac{\delta_{N[i,2]} T_{N[i,2]}}{G_n(T_{N[i,2]})} \frac{\delta_{N[i,1]} T_{N[i,1]}}{G(T_{N[i,1]})} \frac{\delta_{N[i,2]} T_{N[i,2]}}{G(T_{N[i,2]})} \right] \right| \\
&\leq L^2 \frac{1}{n} \sum_{i=1}^{n} \left| \frac{1}{G_n(T_{N[i,1]}) G_n(T_{N[i,2]})} - \frac{1}{G(T_{N[i,1]}) G(T_{N[i,2]})} \right| \\
&\leq L^2 \frac{1}{n} \sum_{i=1}^{n} \frac{1}{G_n(T_{N[i,1]})} \left| \frac{1}{G_n(T_{N[i,2]})} - \frac{1}{G(T_{N[i,2]})} \right| \\
&\quad + L^2 \frac{1}{n} \sum_{i=1}^{n} \frac{1}{G(T_{N[i,2]})} \left| \frac{1}{G_n(T_{N[i,1]})} - \frac{1}{G(T_{N[i,1]})} \right| \\
&\leq L^2 \frac{1}{n} \frac{1}{G_n^2(L) G(L)} \sum_{i=1}^{n} \left| G_n(T_{N[i,2]}) - G(T_{N[i,2]}) \right| \\
&\quad + L^2 \frac{1}{n} \frac{1}{G_n^2(L) G(L)} \sum_{i=1}^{n} \left| G_n(T_{N[i,1]}) - G(T_{N[i,1]}) \right| \\
&\leq 2L^2 \frac{1}{G(L) G_n(L)} \sup_{0 \leq t < \infty} \left| G_n(t) - G(t) \right|
\end{aligned} \tag{5.58}$$

$\to 0 \quad a.s.$ by (5.6) and because of the result $\sup_{0 \leq t \leq T_K} |G_n(t) - G(t)| \to 0$ *a.s.*, due to [35] (compare [22], Theorem 10).
This completes the proof. □

Finally, the following theorem states a rate of convergence for the estimator (5.52).

Theorem 5.7. *Let the assumptions* **(A1)-(A3)** *hold. Let the estimate* $\widetilde{\sigma}^{2\ (NN)}$ *be given by (5.52) with cubic partition of* $\mathbb{R}^d$ *with side length* h_n *of the cubes* ($n \in \mathbb{N}$)*. Moreover, assume the Lipschitz conditions*

$$|m(x) - m(t)| \leq \Gamma \|x - t\|^{\alpha},\ x, t \in \mathbb{R}^d,$$

and

$$|\sigma^2(x) - \sigma^2(t)| \leq \Lambda \|x - t\|^{\beta},\ x, t \in \mathbb{R}^d,$$

($0 < \alpha \leq 1,\ 0 < \beta \leq 1,\ \Gamma,\ \Lambda \in \mathbb{R}_+,\ \|\ \|$ *denoting the Euclidean norm*)*.*
Then, with

$$h_n \sim n^{-\frac{1}{d+2\beta}}$$

one gets

$$\int |\widetilde{\sigma}_n^{2\ (NN)} - \sigma^2(x)|\mu(dx) = O_P\left(\left(\frac{\log n}{n}\right)^{\frac{1}{6}} + \max\left\{n^{-\frac{2\alpha}{d}}, n^{-\frac{\beta}{2\beta+d}}\right\}\right)$$

(Rate of convergence of the partitioning estimator of the local variance under censoring and unknown survival function)

Proof. We note, with (5.53) and (5.46),

$$\begin{aligned}
&\int |\widetilde{\sigma}_n^2(x)^{(NN)} - \sigma^2(x)|\mu(dx)\\
&\leq \int |\widetilde{\sigma}_n^2(x)^{(NN)} - \widehat{\widetilde{\sigma}}_n^2(x)^{(NN)}|\mu(dx) + \int |\widehat{\widetilde{\sigma}}_n^2(x)^{(NN)} - \widehat{\widetilde{\sigma}}_n^2(x)|\mu(dx)\\
&\quad + \int |\widehat{\widetilde{\sigma}}_n^2(x) - \sigma^2(x)|\mu(dx)\\
&\leq A_n + B_n + C_n.
\end{aligned}$$

Now, concerning A_n

$$\begin{aligned}
&\int |\widetilde{\sigma}_n^2(x)^{(NN)} - \widehat{\widetilde{\sigma}}_n^2(x)^{(NN)}|\mu(dx)\\
&= \int \left| \frac{\sum_{i=1}^n H_{i,G_n} 1_{A_n(x)}(X_i)}{\sum_{i=1}^n 1_{A_n(x)}(X_i)} - \frac{\sum_{i=1}^n H_{i,G_n} 1_{A_n(x)}(X_i)}{n\mu(A_n(x))} \right| \mu(dx)\\
&\leq U^{**} \int \sum_{i=1}^n 1_{A_n(x)}(X_i) \left| \frac{1}{\sum_{i=1}^n 1_{A_n(x)}(X_i)} - \frac{1}{n\mu(A_n(x))} \right| \mu(dx)
\end{aligned}$$

(a.s., with a random variable $U^{**} < \infty$, because of (5.57))

$$= O_P\left(n^{-\frac{1}{2}} h_n^{-\frac{d}{2}}\right)$$

by the proof of Theorem 4.3 in [11].
Moreover

$$\begin{aligned} B_n &= \int |\widehat{\widehat{\sigma}}_n^2(x)^{(NN)} - \widehat{\widehat{\sigma}}_n^2(x)|\mu(dx) \\ &\leq \frac{1}{n}\sum_{i=1}^n |H_{i,G_n} - H_i| \\ &\quad \text{(see (5.54))} \\ &\leq \frac{1}{n}\sum_{i=1}^n (P_{n,i} + O_{n,i} + I_{n,i} + U_{n,i}) \\ &\quad \text{(see (5.55))} \end{aligned}$$

Now, concerning $P_{n,i}$,

$$\frac{1}{n}\sum_{i=1}^n P_{n,i} \leq \frac{L}{n}\sum_{i=1}^n \delta_i T_i \left|\frac{1}{G_n(T_i)} - \frac{1}{G(T_i)}\right|$$
$$\leq \frac{L^2}{n}\sum_{i=1}^n \sup_{0\leq t\leq T_i} |G_n(t) - G(t)| = O_P\left(\left(\frac{\log n}{n}\right)^{\frac{1}{6}}\right),$$

the latter by the Cauchy-Schwarz inequality and the proof of Satz 4 in [22]. Finally, concerning $U_{n,i}$, (and similarly, for $O_{n,i}$ and $I_{n,i}$) we note that, instead of (5.58), one also obtains, by [11], Corollary 6.1, with a suitable constant γ_d,

$$\begin{aligned} &\left|\frac{1}{n}\sum_{i=1}^n U_i\right| \\ &= \left|\frac{1}{n}\sum_{i=1}^n \left[\frac{\delta_{N[i,1]}T_{N[i,1]}}{G_n(T_{N[i,1]})}\frac{\delta_{N[i,2]}T_{N[i,2]}}{G_n(T_{N[i,2]})} - \frac{\delta_{N[i,1]}T_{N[i,1]}}{G(T_{N[i,1]})}\frac{\delta_{N[i,2]}T_{N[i,2]}}{G(T_{N[i,2]})}\right]\right| \\ &\leq 2\gamma_d L^2 \frac{1}{n}\frac{1}{G_n^2(L)G(L)}\sum_{i=1}^n |G_n(T_i) - G(T_i)| \\ &\leq 2\gamma_d L^2 \frac{1}{G_n^2(L)G(L)} \sup_{0\leq t\leq T_K} |G_n(t) - G(t)| = O_P\left(\left(\frac{\log n}{n}\right)^{\frac{1}{6}}\right), \end{aligned}$$

(the latter as before).

It remains to give a rate for

$$C_n = \int |\widehat{\widehat{\sigma}}_n^2(x) - \sigma^2(x)|\mu(dx).$$

For that introduce again the expansion

$$\begin{aligned} &\boldsymbol{E}\left\{\int \left|\widehat{\widehat{\sigma}}_n^2(x) - \sigma^2(x)\right| \mu(dx)\right\} \\ &\leq \boldsymbol{E}\left\{\int \left|\widehat{\widehat{\sigma}}_n^2(x) - \boldsymbol{E}\widehat{\widehat{\sigma}}_n^2(x)\right| \mu(dx)\right\} + \int \left|\boldsymbol{E}\widehat{\widehat{\sigma}}_n^2(x) - \sigma^2(x)\right| \mu(dx), \end{aligned}$$

where

$$\boldsymbol{E}\left\{\int \left|\widehat{\widehat{\sigma}}_n^2(x) - \boldsymbol{E}\widehat{\widehat{\sigma}}_n^2(x)\right| \mu(dx)\right\} = O\left(\sqrt{\frac{l_n}{n}}\right) = O\left(\left(n_n^{-d} n^{-1}\right)^{\frac{1}{2}}\right),$$

as in (5.49).
Now,

$$\begin{aligned} &\boldsymbol{E}\widehat{\widehat{\sigma}}_n^2(x) \\ &= \int \frac{\boldsymbol{E}\{W_1 | X_1 = z\} 1_{A_n(x)}(z)}{\mu(A_n(x))} \mu(dz) \\ &\text{(according to (5.50))} \\ &\int \frac{\boldsymbol{E}\{(Y_1 - m(X_1))^2 | X_1 = z\} 1_{A_n(x)}(z)}{\mu(A_n(x))} \mu(dz) \\ &+ \int \frac{\boldsymbol{E}\{(m(X_1) - m(X_{N[1,1]}))(m(X_1) - m(X_{N[1,2]})) | X_1 = z\} 1_{A_n(x)}(z)}{\mu(A_n(x))} \mu(dz) \end{aligned}$$

(as in (5.42)).
Then

$$\begin{aligned} &\int \left|\boldsymbol{E}\widehat{\widehat{\sigma}}_n^2(x) - \sigma^2(x)\right| \mu(dx) \\ &\leq \int \left| \frac{(Y_1 - m(X_1))^2 | X_1 = z\} 1_{A_n(x)}(z)}{\mu(A_n(x))} \mu(dz) - \sigma^2(x) \right| \mu(dx) \\ &+ \int \boldsymbol{E}\left\{(m(X_1) - m(X_{N[1,1]}))(m(X_1) - m(X_{N[1,2]})) | X_1 = z\right\} \\ &\text{(because of } \int \frac{1}{\mu(A_n(z))} 1_{A_n(z)} \mu(dx)\mu(dz) \leq 1) \end{aligned}$$

$$\leq \Lambda h_n + \boldsymbol{E}\left\{(m(X_1) - m(X_{N[1,1]}))(m(X_1) - m(X_{N[1,2]}))\right\}$$
$$\leq \Lambda h_n + \left(\boldsymbol{E}\|X_1 - X_{N[1,1]}\|^{2\alpha}\right)^{\frac{1}{2}} \left(\boldsymbol{E}\|X_1 - X_{N[1,2]}\|^{2\alpha}\right)^{\frac{1}{2}}$$
$$= O\left(h_n + n^{-\frac{2\alpha}{d}}\right)$$

Therefore

$$\boldsymbol{E}\left\{\int \left|\widehat{\widehat{\sigma}}_n^2(x) - \sigma^2(x)\right| \mu(dx)\right\}$$
$$\leq O\left(h_n^{-\frac{d}{2}} n^{-\frac{1}{2}}\right) + O\left(n^{-\frac{2\alpha}{d}} + h_n\right).$$

and finally,

$$\int \left|\widehat{\widehat{\sigma}}_n^2(x) - \sigma^2(x)\right| \mu(dx)$$
$$O_P\left(h_n^{-\frac{d}{2}} n^{-\frac{1}{2}}\right) + O\left(n^{-\frac{2\alpha}{d}} + h_n\right).$$

Now, summarizing

$$\int |\widetilde{\sigma}_n^{2\ (NN)} - \sigma^2(x)|\mu(dx)$$
$$= O_P\left(n^{-\frac{1}{2}} h_n^{-\frac{d}{2}}\right) + O_P\left(\left(\frac{\log n}{n}\right)^{\frac{1}{6}}\right)$$
$$+ O_P\left(h_n^{-\frac{d}{2}} n^{-\frac{1}{2}} + n^{-\frac{2\alpha}{d}} + \left(\frac{\log n}{n}\right)^{\frac{1}{6}}\right)$$
$$= O_P\left(\left(\frac{\log n}{n}\right)^{\frac{1}{6}} + \max\left\{n^{-\frac{2\alpha}{d}}, n^{-\frac{\beta}{2\beta+d}}\right\}\right)$$

and hence the assertion. □

Chapter 6
Simulations

For the purpose of empirical investigation of the performance of local variance estimators in the finite sample case, we conduct simulations based on 200 random samples, each of size $n = 200$, from the one-dimensional model

$$Y_i = m(X_i) + \sigma(X_i)\epsilon_i, \tag{6.1}$$

with $\sigma^2(x) = |x|$, $m(x) = |x|^{1/6}$, where $\{X_i\}$ and $\{\epsilon_i\}$ are two independent sequences of independent random variables, $\epsilon_i \sim \mathcal{N}(0,1)$, $X_i \sim \mathcal{U}[-1,1] = \mu$. This first situation is comprehended by Theorem 4.5 with $d = 1$, $\beta = 1$ and $\alpha = \frac{\beta d}{2(2\beta+d)} = 1/6$, where the local variance estimator is

$$\sigma_n^2(x) := \frac{\sum_{i=1}^n (Y_i - Y_{N[i,1]})(Y_i - Y_{N[i,2]}) 1_{A_n(x)}(X_i)}{\sum_{i=1}^n 1_{A_n(x)}(X_i)}, \quad x \in \mathbb{R}^d,$$

as in (4.19). By Theorem 4.5 this yields the optimal convergence rate $n^{-2/3}$ for the mean integrated squared error.

```
> library(stats)
> library(class)
> rm(list = objects())
> d <- 1
> z <- 0
> resultsFW = vector(mode = "numeric", length = runs)
> n <- 200
> x <- as.matrix(runif(n, min = -1, max = 1), d)
> epsilon <- as.matrix(rnorm(n, mean = 0, sd = abs(x)^(1/2)), d)
> m <- function(x) {
+     abs(x)^(1/6)
+ }
> y <- as.vector(m(x) + epsilon)
> sigma2 <- function(x) {
```

```
+       abs(x)
+ }
```

For each simulated sample the first nearest neighbor is searched in the sample and, after removing it, the second nearest neighbor.

```
> NN1 = function(i) {
+       value1 = 0
+       class = c(1:(n - 1))
+       test = as.matrix(x[i, d])
+       train = as.matrix(x[2:n], d)
+       if (i == 1) {
+           train = as.matrix(x[2:n, d])
+           class = c(2:n)
+       }
+       if (i == n) {
+           train = as.matrix(x[1:(n - 1), d])
+           class = c(1:(n - 1))
+       }
+       if ((i > 1) & (i < n)) {
+           tt = as.matrix(x[1:(i - 1), d])
+           zz = as.matrix(x[(i + 1):n, d])
+           train = rbind(tt, zz)
+           class = c(c(1:(i - 1)), c((i + 1):n))
+       }
+       value1 = as.numeric(as.matrix(knn1(train, test, class)))
+       value1
+ }
> NN2 = function(o) {
+       first <- NN1(o)
+       value2 = 0
+       test2 = as.matrix(x[o, d])
+       if (o == 1) {
+           if (first == 2) {
+               train2 = as.matrix(x[3:n, d])
+               class2 = c(3:n)
+           }
+           if (first == n) {
+               train2 = as.matrix(x[2:(n - 1), d])
+               class2 = c(2:(n - 1))
+           }
+           if (first > 2 && first < n) {
+               tt2 = as.matrix(x[2:(first - 1), d])
+               zz2 = as.matrix(x[(first + 1):n, d])
+               train2 = rbind(tt2, zz2)
+               class2 = c(c(2:(first - 1)), c((first + 1):n))
```

```
+         }
+     }
+     if (o == n) {
+         if (first == (o - 1)) {
+             train2 = as.matrix(x[1:(o - 2), d])
+             class2 = c(1:(o - 2))
+         }
+         if (first == 1) {
+             train2 = as.matrix(x[2:(o - 1), d])
+             class2 = c(2:(o - 1))
+         }
+         if (first >= 2 && first < (o - 2)) {
+             tt2 = as.matrix(x[1:(first - 1), d])
+             zz2 = as.matrix(x[(first + 1):(o - 1), d])
+             train2 = rbind(tt2, zz2)
+             class2 = c(c(1:(first - 1)), c((first + 1):(o - 1)))
+         }
+     }
+     if (o > 1 && o <= (n - 2)) {
+         if (first == (o + 1)) {
+             tt2 = as.matrix(x[1:(o - 1), d])
+             zz2 = as.matrix(x[(first + 1):n])
+             train2 = rbind(tt2, zz2)
+             class2 = c(c(1:(o - 1)), c((first + 1):n))
+         }
+         if (first == n) {
+             tt2 = as.matrix(x[1:(o - 1), d])
+             zz2 = as.matrix(x[(o + 1):(first - 1), d])
+             train2 = rbind(tt2, zz2)
+             class2 = c(c(1:(o - 1)), c((o + 1):(first - 1)))
+         }
+         if (first > (o + 1) && first < n) {
+             tt2 = as.matrix(x[1:(o - 1), d])
+             zz2 = as.matrix(x[(o + 1):(first - 1), d])
+             rr2 = as.matrix(x[(first + 1):n, d])
+             train2 = rbind(tt2, zz2, rr2)
+             class2 = c(c(1:(o - 1)), c((o + 1):(first - 1)),
+                 c((first + 1):n))
+         }
+     }
+     if (first == 1) {
+         if (o == 2) {
+             train2 = as.matrix(x[3:n, d])
+             class2 = c(3:n)
+         }
+         if (o == n) {
+             train2 = as.matrix(x[2:(n - 1), d])
+             class2 = c(2:(n - 1))
+         }
+         if (o > 2 && o < n) {
```

```
+             tt2 = as.matrix(x[2:(o - 1), d])
+             zz2 = as.matrix(x[(o + 1):n, d])
+             train2 = rbind(tt2, zz2)
+             class2 = c(c(2:(o - 1)), c((o + 1):n))
+         }
+     }
+     if (first == n) {
+         if (o == (n - 1)) {
+             train2 = as.matrix(x[1:(n - 2), d])
+             class2 = c(1:(n - 2))
+         }
+         if (o == 1) {
+             train2 = as.matrix(x[2:(first - 1), d])
+             class2 = c(2:(first - 1))
+         }
+         if (o >= 2 && o < (n - 2)) {
+             tt2 = as.matrix(x[1:(o - 1), d])
+             zz2 = as.matrix(x[(o + 1):(first - 1), d])
+             train2 = rbind(tt2, zz2)
+             class2 = c(c(1:(o - 1)), c((o + 1):(first - 1)))
+         }
+     }
+     if (first > 1 && first <= (n - 2)) {
+         if (o == (first + 1)) {
+             tt2 = as.matrix(x[1:(first - 1), d])
+             zz2 = as.matrix(x[(o + 1):n])
+             train2 = rbind(tt2, zz2)
+             class2 = c(c(1:(first - 1)), c((o + 1):n))
+         }
+         if (o == n) {
+             tt2 = as.matrix(x[1:(first - 1), d])
+             zz2 = as.matrix(x[(first + 1):(o - 1), d])
+             train2 = rbind(tt2, zz2)
+             class2 = c(c(1:(first - 1)), c((first + 1):(o - 1)))
+         }
+         if (o > (first + 1) && o < n) {
+             tt2 = as.matrix(x[1:(first - 1), d])
+             zz2 = as.matrix(x[(first + 1):(o - 1), d])
+             rr2 = as.matrix(x[(o + 1):n, d])
+             train2 = rbind(tt2, zz2, rr2)
+             class2 = c(c(1:(first - 1)), c((first + 1):(o - 1)),
+                 c((o + 1):n))
+         }
+     }
+     value2 = as.numeric(as.matrix(knn1(train2, test2, class2)))
+     return(value2)
+ }
> xy <- cbind(x, y)
> yNN1 = function(i) {
+     return(xy[NN1(i), d + 1])
```

```
+ }
> yNN2 = function(i) {
+     return(xy[NN2(i), d + 1])
+ }
```

According to Theorem 4.5 we use partitioning by intervals of the length $Cn^{-1/(2\beta+1)} = Cn^{-1/3}$ and discuss various results depending on the choice of this constant C.
First, we choose $C = 1/2$, getting 24 partitioning intervals covering $[-1, 1]$.

```
> C <- 1/2
> cubuslength <- C * n^(-1/(2 + d))
> part <- vector(mode = "numeric", length = 25)
> part[1] = -1
> for (r in 1:25) {
+     part[r + 1] = part[r] + cubuslength
+ }
> Wninum <- function(i, smallx) {
+     value3 = 0
+     Anx = vector(mode = "numeric", length = 2)
+     for (l in (1:25)) {
+         if (smallx > part[l] && smallx < part[l + 1]) {
+             Anx[1] <- part[l]
+             Anx[2] <- part[l + 1]
+         }
+         if (x[i, d] > Anx[1] && x[i, d] < Anx[2])
+             value3 = 1
+         if (x[i, d] < Anx[1] && x > Anx[2])
+             value3 = 0
+         if (smallx < part[l] && smallx > part[l + 1])
+             value3 = 0
+     }
+     return(value3)
+ }
> Wniden <- function(smallx) {
+     Wnidenom = 0
+     for (p in (1:n)) {
+         Wnidenom <- Wnidenom + Wninum(p, smallx)
+     }
+     return(Wnidenom)
+ }
```

Then, the estimator (4.19) was implemented and the performance of the estimator is evaluated by an empirical L_2 error

$$L_2 = \frac{1}{N}\sum_{i=1}^{N} |\sigma_n^2(x_i) - \sigma^2(x_i)|^2, \tag{6.2}$$

being a good approximation of $\int_{-1}^{1} |\sigma_n^2(x) - \sigma^2(x)|^2 \mu(dx) = \frac{1}{2}\int_{-1}^{1} |\sigma_n^2(x) - \sigma^2(x)|^2 dx$, where $\{x_i,\ i = 1, \ldots, N\}$ (N=101) are equidistant grid points on $[-1, 1]$, with $x_{i+1} - x_i = 0.02$.
The results are saved.

```
> sigma2nFW <- function(smallx) {
+       sumi = 0
+       {
+           for (i in (1:n)) sumi <- (y[i] - yNN1(i)) * (y[i] - yNN2(i)) *
+               Wninum(i, smallx) + sumi
+       }
+       if (Wniden(smallx) != 0)
+           estimator <- sumi/Wniden(smallx)
+       else estimator <- 0
+       return(estimator)
+ }
> L2eFWvorsumi <- function(smallx) {
+       (abs(sigma2nFW(smallx) - sigma2(smallx))^2)
+ }
> for (smallx in seq(from = -1, to = 1, by = 0.02)) {
+       L2eFW <- sum(L2eFWvorsumi(smallx))/101
+ }
> z <- z + 1
> resultsFW[z] = L2eFW
```

In order to compare estimators under similar assumptions, the second simulated situation extends the results of Theorem 3.3 for

$$|m(x) - m(z)| \leq C\|x - z\|^{\alpha}, \quad x, z \in \mathbb{R}^d,$$

and

$$|\sigma^2(x) - \sigma^2(z)| \leq D\|x - z\|^{\beta}, \quad x, z \in \mathbb{R}^d,$$

($\|\ \|$ denoting the Euclidean norm).
In particular, with $d = 1$, $\beta = 1$ and $\alpha = \frac{\beta d}{2(2\beta+d)}$ we have Hölder continuous m, ($\alpha = 1/6$) and Lipschitz continuous σ^2.
The estimator is given by

$$\sigma_n^2(x)^{(LA)} = \sum_{i=1}^{n} W_{n,i}^{(1)}(x) \cdot \left(Y_i^2 - \left(\sum_{i=1}^{n} W_{n,i}^{(2)}(X_i) Y_i \right)^2 \right),$$

as in (3.10), with partitioning weights

$$W_{n,i}^{(l)}(x) = \frac{1_{A_n^{(l)}(x)}(X_i)}{\sum_{i=1}^{n} 1_{A_n^{(l)}(x)}(X_i)},\ l = 1, 2.$$

The partitioning intervals $A_{n,j}^{(1)}$ are of length $C_1 n^{-1/(2\beta+1)} = C_1 n^{-1/3}$, $C_1 = 1/2$, $j = 1, \ldots, 24$. In our comparative simulation we use the foregoing number 24 of intervals for estimating m as well as for finally estimating σ^2. Thus in the ansatz $C_2 n^{-1/(2\alpha+1)} = C_2 n^{-3/4}$ ($\alpha = 1/6$) for the length of partitioning intervals $A_{n,j}^{(2)}$, which is of optimal order for the isolated problem of estimating the Hölder continuous function m, one has $C_2 = 4.43$.

```
> d <- 1
> w <- 0
> resultsPI = vector(mode = "numeric", length = runs)
> n <- 200
> x <- as.matrix(runif(n, min = -1, max = 1), d)
> epsilon <- as.matrix(rnorm(n, mean = 0, sd = abs(x)^(1/2)), d)
> m <- function(x) {
+     abs(x)^(1/6)
+ }
> y <- as.vector(m(x) + epsilon)
> sigma2 <- function(x) {
+     abs(x)
+ }
> C1 <- 1/2
> cubuslength1 <- C1 * n^(-1/(2 + d))
> part1 <- vector(mode = "numeric", length = 25)
> part1[1] = -1
> for (r in 1:25) {
+     part1[r + 1] = part1[r] + cubuslength1
+ }
> Wninum1 <- function(i, smallx) {
+     value3 = 0
+     Anx1 = vector(mode = "numeric", length = 2)
+     for (l in (1:25)) {
+         if (smallx > part1[l] && smallx < part1[l + 1]) {
+             Anx1[1] <- part1[l]
+             Anx1[2] <- part1[l + 1]
+         }
+         if (x[i, d] > Anx1[1] && x[i, d] < Anx1[2])
+             value3 = 1
+         if (x[i, d] < Anx1[1] && x > Anx1[2])
+             value3 = 0
+         if (smallx < part1[l] && smallx > part1[l + 1])
+             value3 = 0
+     }
+     return(value3)
+ }
> Wniden1 <- function(smallx) {
+     Wnidenom1 = 0
+     for (p in (1:n)) {
+         Wnidenom1 <- Wnidenom1 + Wninum1(p, smallx)
+     }
+     return(Wnidenom1)
```

```
+ }
> C2 <- 4.43
> alpha <- d/(2 * (2 + d))
> cubuslength2 <- C2 * n^(-1/(1 + 2 * alpha))
> part2 <- vector(mode = "numeric", length = 25)
> part2[1] = -1
> for (r in 1:25) {
+     part2[r + 1] = part2[r] + cubuslength2
+ }
> Wninum2 <- function(i, smallx) {
+     value3 = 0
+     Anx2 = vector(mode = "numeric", length = 2)
+     for (l in (1:25)) {
+         if (smallx > part2[l] && smallx < part2[l + 1]) {
+             Anx2[1] <- part2[l]
+             Anx2[2] <- part2[l + 1]
+         }
+         if (x[i, d] > Anx2[1] && x[i, d] < Anx2[2])
+             value3 = 1
+         if (x[i, d] < Anx2[1] && x > Anx2[2])
+             value3 = 0
+         if (smallx < part2[l] && smallx > part2[l + 1])
+             value3 = 0
+     }
+     return(value3)
+ }
> Wniden2 <- function(smallx) {
+     Wnidenom2 = 0
+     for (p in (1:n)) {
+         Wnidenom2 <- Wnidenom2 + Wninum2(p, smallx)
+     }
+     return(Wnidenom2)
+ }
> mnPI <- function(smallx) {
+     sumi = 0
+     for (i in (1:n)) {
+         sumi <- (y[i] * Wninum2(i, smallx) + sumi)
+     }
+     if (Wniden2(smallx) != 0)
+         estimator2 <- sumi/Wniden2(smallx)
+     else estimator2 <- 0
+     return(estimator2)
+ }
```

Then, the local variance estimation is implemented

```
> sigma2nPI <- function(smallx) {
+     summi = 0
+     for (i in (1:n)) {
+         summi <- (y[i] - mnPI(smallx))^2 * Wninum1(i, smallx) +
```

```
+               summi
+       }
+       if (Wniden1(smallx) != 0)
+           estimator <- summi/Wniden1(smallx)
+       else estimator <- 0
+       return(estimator)
+ }
```

The performance of the estimator is evaluated by an empirical L_2 error as in (6.2)

```
> L2ePIvorsumi <- function(smallx) {
+       (abs(sigma2nPI(smallx) - sigma2(smallx))^2)
+ }
> for (smallx in seq(from = -1, to = 1, by = 0.02)) {
+       L2ePI <- sum(L2ePIvorsumi(smallx))/101
+ }
> w <- w + 1
> resultsPI[w] = L2ePI
```

The performance of the two estimators (4.19) and (3.10) was compared and represented by a boxplot (Figure 6.1, left-hand and right-hand side, respectively).

Comparing the two estimation methods on the basis of simulations, it is first of all important to notice that there is a remarkable computational cost difference. The estimation method (4.19) needs a lower computational time. Second, as can be seen in Figure 6.1, this estimation performs better than the plug-in version, using each time the same number 24 of partitioning intervals. The empirical L_2 error for the estimator (4.19) shows a lower first quartile and lower second quartile (median).
The question how to choose a proper partition of the support motivated us to compare the empirical L_2 error of the estimator (4.19) for different choices of the partition, i.e., besides the already discussed case $C = 1/2$ (24 partitioning intervals) also for $C = 1/4$ (47 partitioning intervals), $C = 1$ (12 partitioning intervals), $C = 2$ (6 partitioning intervals). The boxplot in Figure 6.2 compares these four cases. Obviously, a large C leads to the trivial case of 1 partitioning interval $[-1, 1]$ with large median of the empirical L_2-error. A classical way of choosing the parameter C is using an independent validation sample.
Further we simulated the case $d = 2$ extending the model (6.1), with X_i uniformly distributed on $[-1, 1] \times [-1, 1]$, $\sigma^2 : \mathbb{R}^2 \to \mathbb{R}$, $\sigma^2(\mathbf{x}) = \|\mathbf{x}\|$ and $m : \mathbb{R}^2 \to \mathbb{R}$, $m(\mathbf{x}) = \|\mathbf{x}\|^{1/4}$, where $\| \ \|$ denotes the Euclidean norm. The situation is comprehended by the Theorem 4.5 with $d = 2$, $\beta = 1$,

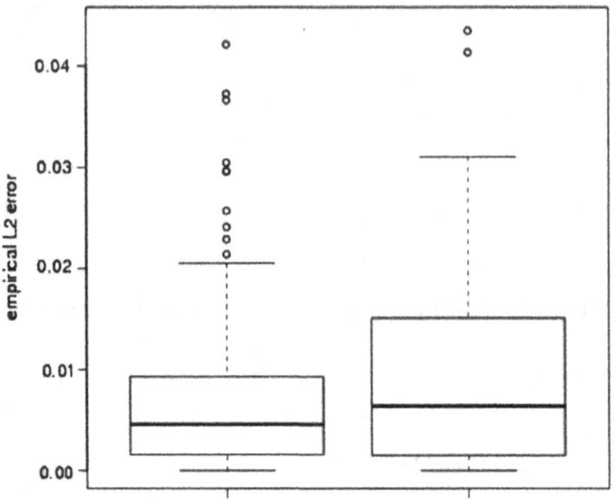

Fig. 6.1 Boxplot of the L_2 error for the estimator (4.19) (on the left-hand side) and the estimator (3.10) (on the right-hand side)

$\alpha = \frac{\beta d}{2(2\beta+d)} = 1/4$, giving with $h_n \sim n^{-1/4}$ the optimal convergence rate $n^{-1/2}$.
We chose $C = 1.25$ in order to get 36 partitioning squares, corresponding to the 6 partitioning intervals in the favorite case of Figure 2 for $d = 1$. For 200 runs of sample size 200 the empirical L_2-error was calculated according to (6.2) and represented in a boxplot in Figure 6.3.
The slightly modified codex for the case $d = 2$ is not reported.

Remark 6.1. The boxplots are generated by the R function **boxplot** under default parameter range=1.5. It means that the lines ("whiskers") extend to the largest/smallest observation that falls within a distance of 1.5 times the interquartile difference. Any data not included between the whiskers are plotted as outliers.
We also represented the boxplots under the untypical value of zero for the range, causing that the whiskers extend to the data extremes and no observation is considered as outlier.
However, the trends illustrated in all pictures remain the same.

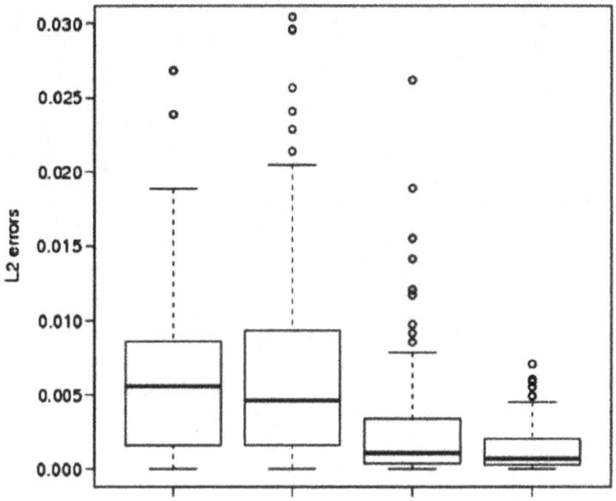

Fig. 6.2 Boxplot of the L_2 error for the estimator (4.19) for $C = 1/4$, $C = 1/2$, $C = 1$ and $C = 2$, respectively

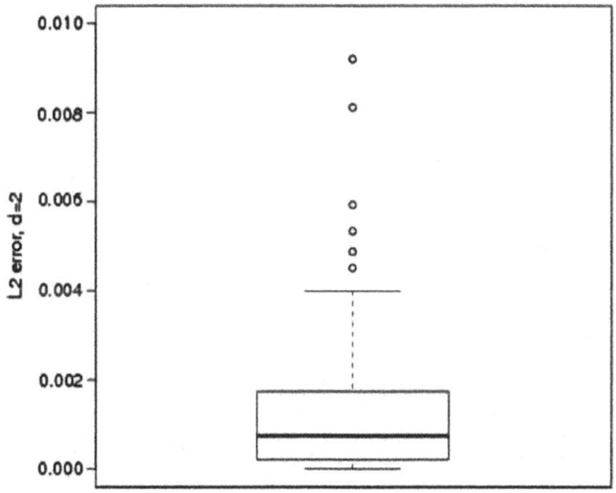

Fig. 6.3 Boxplot of the L_2 error for the estimator (4.19) for $C = 1.25$, $d = 2$

References

1. L.D. Brown and M. Levine. Variance estimation in nonparametric regression via the difference sequence method. *Annals of Statistics*, 35:2219–2232, 2007.
2. T. Cai, M. Levine, and L. Wang. Variance function estimation in multivariate nonparametric regression with fixed design. *Journal of Multivariate Analysis*, 100:126–136, 2009.
3. C. De Boor. *A Practical Guide to Splines.* Springer, New York, 1978.
4. L. Devroye, D. Schäfer, L. Györfi, and H. Walk. The estimation problem of minimum mean squared error. *Statistics & Decisions*, 21:15–28, 2003.
5. N. Dunford and J.T. Schwartz. *Linear Operators, General Theory.* Wiley Classics Library, New York, 1958.
6. D. Evans. Estimating the variance of multiplicative noise. *18th International Conference on Noise and Fluctuations, ICNF, in AIP Conference Proceedings*, 780:99–102, 2005.
7. D. Evans and A.J. Jones. Non-parametric estimation of residual moments and covariance. *Proceedings of the Royal Society A*, 464:2831–2846, 2008.
8. J. Fan and I. Gijbels. Censored regression: local linear approximations and their applications. *Journal of the American Statistical Association*, 89(426):560–570, 1994.
9. J. Fan and Q. Yao. Efficient estimation of conditional variance functions in stochastic regression. *Biometrika*, 85:645–660, 1998.
10. L. Györfi. Universal consistencies of a regression estimate for unbounded regression functions. In *Nonparametric Functional Estimation and Related Topics*, Ed. G. Roussas, Kluwer Academic Publisher, pages 329–338, 1991.
11. L. Györfi, M. Kohler, A. Krzyżak, and H. Walk. *A Distribution-Free Theory of Nonparametric Regression.* Springer, New York, 2002.
12. P. Hall and P.J. Carroll. Variance function estimation in regression: The effect of estimating the mean. *Journal of the Royal Statistical Society, Ser. B*, 51:3–14, 1989.
13. P. Hall, J.W. Kay, and D.M. Titterington. Asymptotically optimal difference-based estimation of variance in nonparametric regression. *Biometrika*, 77(3):521–528, 1990.
14. W. Härdle and A. Tsybakov. Local polynomial estimators of the volatility function in nonparametric autoregression. *Journal of Econometrics*, 81:223–242, 1997.

15. E. L. Kaplan and P. Meier. Nonparametric estimation from incomplete observations. *Journal of the American Statistical Association*, 53:457–481, 1958.
16. M. Kohler. Nonparametric regression with additional measurement errors in the dependent variable. *Journal of Statistical Planning and Inference*, 136:3339–3361, 2006.
17. M. Kohler, A. Krzyżak, and H. Walk. Rates of convergence for partitioning and nearest neighbor regression estimates with unbuonded data. *Journal of Mutlivariate Analysis*, 97:311–323, 2006.
18. E. Liitiäinen, F. Corona, and A. Lendasse. Non-parametric residual variance estimation in supervised learning. *IWANN'07 Proceedings of the 9th International Work-Conference on Artificial Neural Networks. Lecture Notes in Computer Science: Computational and Ambient Intelligence*, 4507:63–71, 2007.
19. E. Liitiäinen, F. Corona, and A. Lendasse. On nonparametric residual variance estimation. *Neural Processing Letters*, 28:155–167, 2008.
20. E. Liitiäinen, F. Corona, and A. Lendasse. Residual variance estimation using a nearest neighbor statistic. *Journal of Multivariate Analysis*, 101:811–823, 2010.
21. M. Loève. *Probability Theory*. 4th ed. Springer, Berlin, 1977.
22. K. Mathe. *Regressionanalyse mit zensierten Daten, PhD Thesis*. Institute of Stochastics and Applications, Universität Stuttgart, 2006.
23. H.G. Müller and U. Stadtmüller. Estimation of heteroscedasticity in regression analysis. *Annals of Statistics*, 15:610–625, 1987.
24. H.G. Müller and U. Stadtmüller. On variance function estimation with quadratic forms. *Journal of Statistical Planning and Inference*, 35:213–231, 1993.
25. U. Müller, A. Schick, and W. Wefelmeyer. Estimating the error variance in nonparametric regression by a covariate-matched u-statistic. *Statistics*, 37(3):179–188, 2003.
26. A. Munk, N. Bissantz, T. Wagner, and G. Freitag. On difference based variance estimation in nonparametric regression when the covariate is high dimensional. *Journal of the Royal Statistical Society, Ser. B*, 67:19–41, 2005.
27. M.H. Neumann. Fully data-driven nonparametric variance estimators. *Statistics*, 25:189–212, 1994.
28. Z. Pan and X. Wang. A wavelet-based nonparametric estimator of the variance function. *Computational Economics*, 15:79–87, 2000.
29. D. Ruppert, M.P. Wand, U. Holst, and O. Hössjer. Local polynomial variance-function estimation. *Technometrics*, 39(3):262–273, 1997.
30. V. Spokoiny. Variance estimation for high-dimensional regression models. *Journal of Multivariate Analysis*, 82:111–133, 2002.
31. U. Stadtmüller and A.B. Tsybakov. Nonparametric recursive variance estimation. *Statistics*, 27:55–63, 1995.
32. J.M. Steele. An Efron-Stein inequality for nonsymmetric statistics. *Annals of Statistics*, 14:753–758, 1986.
33. W. Stute and J.-L. Wang. The strong law under censorship. *Annals of statistics*, 21:1591–1607, 1993.
34. H. Walk. Strong laws of large numbers and nonparametric estimation. In *Recent Developments in Applied Probability and Statistics*. L. Devroye, B. Karasözen, M. Kohler, R. Korn, eds., pages 183-214. Physica-Verlag, Heidelberg, 2010.
35. L. Wang, L.D. Brown, T. Cai, and M. Levine. Effect of mean on variance function estimation in nonparametric regression. *Annals of Statistics*, 36:646–664, 2008.